Tree-Based Heterogeneous FPGA Architectures

Umer Farooq · Zied Marrakchi
Habib Mehrez

Tree-Based Heterogeneous FPGA Architectures

Application Specific Exploration and Optimization

Umer Farooq
COMSATS Institute of Information Technology
Kamboh House
Syed Chiragh Shah Town
Kasur 55050
Pakistan

Habib Mehrez
Paris VI
Laboratoire LIP6
Université Pierre et Marie Curie
4, Place Jussieu
75252 Paris
France

Zied Marrakchi
FLEXRAS Technologies
Boulevard Anatole
Tour Pleyel Ouest 153
93521 Saint Denis
France

ISBN 978-1-4899-9612-1 ISBN 978-1-4614-3594-5 (eBook)
DOI 10.1007/978-1-4614-3594-5
Springer New York Heidelberg Dordrecht London

Softcover reprint of the hardcover 1st edition 2012

Printed on acid-free paper

Springer is part of Springer Science+Business Media (www.springer.com)

Dedicated to our families

Preface

The generalized and programmable nature of Field Programmable Gate Arrays (FPGAs) has made them a popular choice for the implementation of digital circuits. However, the programmability of FPGAs makes them larger, slower, and more power consuming than their counterpart ASICs; hence making them unsuitable for applications requiring high density, performance, and low power consumption. The main theme of this work is to improve the area of FPGAs. For this purpose, a detailed exploration and optimization of two FPGA architectures is performed: one is the well-known mesh-based FPGA architecture while the other is tree-based architecture that remains relatively unexplored despite its better performance and routing predictability. Further, a detailed comparison between the two architectures is presented to highlight their respective advantages and disadvantages.

The exploration and optimization of two architectures start with the introduction of heterogeneous hard-blocks in both architectures. In this work, first we present a new environment for exploration of tree-based heterogeneous FPGA architecture. This environment is flexible in nature and allows to explore different architecture techniques with varying types of hard-blocks. Further, in this work, we present an exploration environment for mesh-based heterogeneous FPGA architecture. The two environments are used to explore a number of techniques for both architectures. These techniques are later evaluated using different heterogeneous benchmarks that are placed and routed on the two architectures using a specifically developed software flow. A detailed comparison between different techniques of the two architectures is performed and results show that on average, tree-based architecture gives better overall results than mesh-based architecture.

Generalized mesh and tree-based FPGA architectures are further improved by turning them into application-specific FPGAs. An application-specific inflexible FPGA (ASIF) is a modified FPGA with reduced flexibility and improved density. This work initially presents a new tree-based homogeneous ASIF and when compared to an equivalent tree-based FPGA, it gives 64% area gain. Further, the comparison between equivalent mesh and tree-based ASIFs shows that tree-based ASIF gives 12% better area results than mesh-based ASIF. We also extend the

ASIF to the heterogeneous domain and experimental results show that, on average, tree-based heterogeneous ASIF gives 70% area gain when compared to equivalent tree-based heterogeneous FPGA. Further, the comparison between heterogeneous mesh and tree-based ASIFs reveals that tree-based ASIF gives either equal or better results than mesh-based ASIF.

Acknowledgments

We would like to take this opportunity to express our sincere thanks to the Director of System on Chip Lab, Professor Alix Munier KORDON, for her efforts which provided a wonderful work environment at SoC-LIP6.

We are extremely thankful to Mr. Daniel CHILLET, Associate Professor at IRISA Lannion, Mr. Laurent FESQUET, Associate Professor at TIMA, Grenoble, Professor Francois ANCEAU, Full Professor at LIP6/University Pierre & Marie Curie, Professor Jean-Luc Danger, Full Professor at ENST/Telecom ParisTech, and Professor Bertrand GRANADO, Full Professor at ENSEA, Cergy Pontoise for reviewing this work. We are particularly thankful to Professor Christian Masson who perceived the importance of this work at its preliminary stage and gave very useful insights for its quality improvement.

We express our sincere gratitude and deep appreciation to Dr. Husain Parvez and Dr. Hayder Mrabet for providing continued technical support. This book would not have been possible without their support and guidance. We would like to thank our colleagues at SoC-LIP6 and the FPGA team colleagues Emna Amouri and Alp Kilic.

Paris, France, February 2012

Umer Farooq
Zied Marrakchi
Habib Mehrez

Contents

Acronyms

ALM	Adaptive logic module
ASIC	Application specific integrated circuit
ASIF	Application specific inflexible FPGA
ASIF-EPER	ASIF efficient partition/placement efficient routing
ASIF-EPNR	ASIF efficient partition/placement normal routing
ASIF-IPER	ASIF ideal partition/placement efficient routing
ASIF-IPIR	ASIF ideal partition/placement ideal routing
ASIF-IPNR	ASIF ideal partition/placement normal routing
ASIF-NPER	ASIF normal partition/placement efficient routing
ASIF-NPNR	ASIF normal partition/placement normal routing
BLE	Basic logic element
BLIF	Berkeley logic interchange format
CAD	Computer aided design
cASIC	Configurable application specific integrated circuit
CLB	Configurable logic block
CMOS	Complementary metal oxide semiconductor
DMSB	Downward mini switch box
DSP	Digital signal processing
FFT	Fast fourier transform
FIR	Finite impulse response
FPGA	Field programmable gate array
HB	Hard block
HDL	Hardware description language
IIR	Infinite impulse response
LAB	Logic array block
LB	Logic block
LUT	Look-up table
MAC	Multiply accumulate
MCNC	Microelectronic center of North Carolina
MLAB	Memory logic array block
MRAM	Magnetic random access memory

NRE	Non-recurring engineering
PLL	Phase locked loop
RAM	Random access memory
RTL	Register transfer level
SRAM	Static random access memory
UMSB	Upward mini switch box
VHSIC	Very high speed integrated circuit
VHDL	VHSIC hardware description language
VPR	Versatile place & route

Chapter 1
Introduction

1.1 Background

Field Programmable Gate Arrays (FPGAs) are pre-fabricated silicon devices that can be electrically programmed to become almost any kind of digital circuit or system. First modern era FPGA was introduced almost two and a half decades ago. That FPGA contained very small number of logic blocks and I/Os. Since then, FPGAs have witnessed an enormous expansion both in terms of capacity and market. They have come a long way from the devices that were once considered only as glue logic to the devices that can now implement complete applications. FPGAs are now widely used for implementing digital circuits in a wide variety of markets including telecommunications, automotive systems and consumer electronics.

FPGAs consist of an array of blocks of potentially different types, including general purpose logic blocks and specific purpose hard blocks like memory and multiplier blocks. Among these blocks, general purpose logic blocks are programmable and along with specific purpose hard blocks they are surrounded by a programmable routing fabric that allows these blocks to be programmably interconnected. The array of blocks along with routing fabric is surrounded by programmable input/output blocks that connect the chip to the outside world. The "programmable" term in FPGA indicates that virtually any hardware function can be programmed into it after its fabrication. This customization is realized with the help of programming technology, which is a method that changes the behavior of the chip in the field after its fabrication.

FPGAs rely on an underlying programming technology that is used to control the programmable switches that give FPGAs their programmability. There are a number of programming technologies and their differences have a significant impact on programmable logic architecture. Earlier programmable logic devices used very small fuses as their programming technology [55]. However, later this programmable technology was replaced by now widely used static memory based programming technology. Most commercial vendors [76, 126] use Static Random Access Memory (SRAM) based programming technology because of its easy re-programmability and

U. Farooq et al., *Tree-Based Heterogeneous FPGA Architectures*,
DOI: 10.1007/978-1-4614-3594-5_1,

the use of standard CMOS process technology. Although, some other programming technologies like flash [2] and anti-fuse [23] are both smaller in area and are non-volatile, the use of standard CMOS manufacturing process makes the SRAM-based programming technology dominating. As a result SRAM-based FPGAs can use the latest CMOS technology and therefore benefit from increased integration, higher speed and lower dynamic power consumption of new process with smaller geometry. Primarily, in an FPGA, SRAM cells are used to program multiplexors that steer the interconnect of FPGAs. Further, they are also used to store the data in general purpose logic blocks also termed as Configurable Logic Blocks (CLBs) that are typically used in SRAM based FPGAs to implement logic functions.

The flexibility and reprogrammability of FPGAs leads to lower Non-Recurring Engineering (NRE) cost and faster time to market than more customized approaches such as Application Specific Integrated Circuit (ASIC) design. The pre-fabricated and programmable nature of FPGAs provides digital circuit designers access to the benefits of latest process technology. In case of custom design, however, significant time and money must be spent on ever-increasing complex issues associated with design and fabrication using latest custom VLSI process technology. On contrary, FPGA-based design cycle time and NRE cost is much lower than full-custom or standard-cell based ASIC layouts.

FPGAs pay for these advantages, however, with some significant disadvantages. Compared with the non-programmable devices, FPGAs have higher area, lower performance and higher power consumption. The large area gap affects the costs, and also limits the size of the designs that can be implemented on FPGAs. The loss in performance also drives up costs as more parallelism and hence greater area may be needed to achieve a performance target. Also it simply may not be possible to achieve the desired performance on an FPGA. Similarly, higher power consumption often limits FPGAs from markets requiring high efficiency in terms of power consumption. Together, this area, performance and power gap limits the applicability of FPGAs when area, speed and/or power requirements of an application are not met. Authors in [60] have reported that FPGAs are 20–35 times larger, 3–4 times slower and 7–14 times more power consuming than ASICs. As a result of this large area, performance and power gap between FPGAs and ASICs, FPGAs become unsuitable for some applications. To address this limitation, a range of alternatives to FPGAs exist.

The primary alternative to an FPGA is an ASIC that has speed, power and area advantages over an FPGA. However, compared to FPGAs, ASICs have certain disadvantages in the form of higher non-recurring engineering (NRE) cost, longer manufacturing time and increasingly complicated design process. While an ASIC implementation offers area, performance and power gains, the difficulties associated to their design process have led to the development of devices that lie in between FPGAs and ASICs. These devices are termed as Structured-ASICs. Structured-ASICs can cut the NRE cost of ASICs by more than 90% while speeding up significantly their time to market [125]. Structured-ASICs contain array of optimized elements which implement a desired functionality by making changes to few upper mask layers. The density and performance of a Structured-ASIC is directly related to the number of

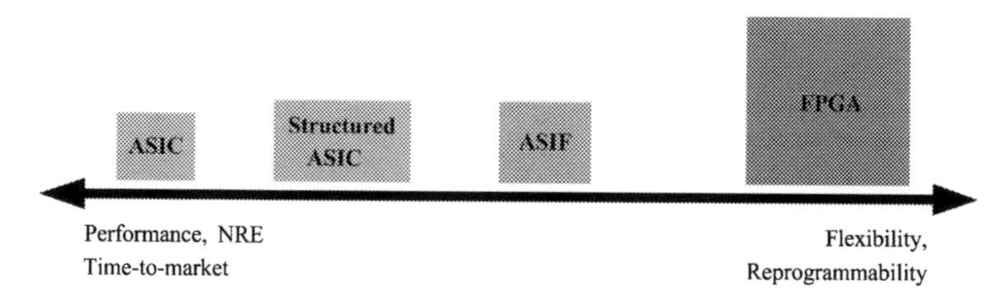

Fig. 1.1 Comparison of different platforms used for implementing digital applications

mask layers that are available for customization. Structured-ASICs are explored or manufactured by several companies [41, 91, 103, 125]. Although Structured-ASICs give lower NRE cost as compared to the standard cell ASICs, their overall efficiency is not as good as that of ASICs and additional cost of an ASIC implementation is not always prohibitive.

FPGA vendors have also started giving provision to migrate FPGA based application to Structured-ASIC. In this regard, Altera has proposed a clean migration methodology [98] that ensures equivalence verification between FPGA and its Structured-ASIC (known as HardCopy [56]). However, migration of an FPGA based application to HardCopy can execute only a single circuit. An Application Specific Inflexible FPGA (ASIF) [93], on the other hand, comprises of optimized logic and routing resources like Structured-ASIC but retains enough flexibility to implement a set of pre-determined applications that operate at mutually exclusive times. Contrary to Structured-ASIC which is basically a modified form of ASIC and which is capable of implementing only one application, an ASIF is a modified form of an FPGA and it can implement a set of application for whom it is designed. However, unlike FPGAs that are generalized in nature, an ASIF contains more customized logic and routing resources and it has only enough flexibility that is required to implement a predetermined set of applications. Figure 1.1 presents a rough comparison of different platforms that can be used for implementing digital applications.

1.2 Book Motivation and Contributions

In general, the overall efficiency of an FPGA is in inverse relation with its flexibility i.e. improvement in one aspect causes a deterioration in the other and vice versa. The main theme of this work is to remedy the drawbacks that are associated with FPGAs with/without compromising their advantages. For this purpose, we explore and optimize a relatively new and unexplored tree-based (hierarchical) architecture along with an established and well investigated mesh-based (island-style) architecture. Although the two architectures comprise of similar logic and routing resources, it is the arrangement of these resources that converts them into altogether different architectures exhibiting different area, performance and power results. In a tree-based architecture, logic resources are arranged in clusters and these clusters are connected to each other recursively to form a hierarchical structure. On the other hand, logic resources in a mesh-based architecture are arranged in an island-style and these resources are connected to each other using uniform routing resources that surround them. In order to explore the two architectures, a new exploration environment for tree-based architecture and an optimized, enhanced environment for mesh-based architecture are used. The two environments are based on a mixture of generalized and specifically developed tools for mapping different applications on the two architectures.

While exploring and optimizing two architectures, our main emphasis is on the area optimization; performance and power optimization are not performed in this work. Area improvement generally implies smaller architectures which result in an improvement both in performance and power consumption. In order to improve the area of the two FPGA architectures, following two broad techniques are employed:

1. Improve the utilization of logic resources of the architecture.
2. Improve the utilization of routing resources of the architecture.

Classic FPGA architectures used only a single type of block that provided the basic logic capability for the implementation of almost any kind of application. Although the use of logic blocks makes FPGAs a good alternative for almost any kind of application, it requires a large amount of logic and routing resources. Now a days, a lot of DSP and arithmetic intensive applications use memories, adders and multiply operations. When these applications are mapped on FPGAs, considerable amount of logic and routing resources can be saved in FPGAs by mapping such operations directly on the specific hard-blocks that are embedded in the architecture along with logic blocks. By embedding hard-blocks directly into the architecture, the overall size of the architecture can be reduced which eventually results in improved area and performance results. The types and quantities of hard-blocks in an FPGA can be decided from the application domain for which an FPGA is required. In this work, we embed hard-blocks in the two architectures under consideration to reduce the overall architecture size and hence improve the utilization of logic resources of the architecture.

The area of an FPGA can be further decreased by optimizing the routing network of an FPGA for a given set of application circuits. By optimization, here, we mean that routing network has a reduced flexibility and it can implement only a pre-determined set of applications. Such a reduced FPGA is called here as an Application Specific Inflexible FPGA (ASIF). An ASIF can be either used in the same scenario as that of Structured-ASIC where a product is initially designed and tested on an FPGA and later it is migrated to an ASIF for high volume production. However, the second and major application of ASIF can be a product that performs different tasks at different times. Such a product may comprise of a video application, a multi-standard radio application, or any set of DSP functionalities required at different times. For example, in the case of a camera different encoders and decoders are required for video and image processing. Further, various compression techniques can be used both for images (e.g. JPEG and PNG etc) and video recording (e.g. MPEG-4 and

H.264). So different digital circuits can be designed and tested on an FPGA and later, for high volume production, the FPGA can be reduced to an ASIF for the given application circuits. So in this work we improve the area of the FPGA architectures by first efficiently incorporating hard-blocks in them and then optimizing their routing networks for a particular set of applications.

The major contributions of this book are as follows:

1.2.1 Exploration Environment for Heterogeneous Tree-Based FPGA Architectures

This work presents a new exploration environment for tree-based heterogeneous FPGA architecture. This environment is generalized and flexible in nature and can be used to explore different architectural topologies with a varying range of logic blocks and hard-blocks. Further, this work also presents an exploration environment for mesh-based heterogeneous FPGA architectures. The environments of two architectures are used to explore and evaluate a number of techniques for both architectures.

The exploration and evaluation of two architectures start with respective architecture definition where separate architecture description mechanisms are used to select different architecture parameters for the two architectures under consideration. Once the architectures are defined, separate software CAD flows are then used to map application circuits on the two architectures. Each software flow uses appropriate techniques to optimize respective architecture. Although, the main objective of the book is not to establish the supremacy of one architecture over the other, however, a detailed comparison between mesh-based and tree-based architectures is presented using 21 heterogeneous benchmarks. Comparison results reveal that tree-based heterogeneous FPGA architecture gives better overall results than mesh-based heterogeneous FPGA architecture.

1.2.2 Exploration of Tree-Based ASIF Architecture

An Application Specific Inflexible FPGA (ASIF) is a modified form of FPGA with reduced flexibility that can implement a set of application circuits which will operate at mutually exclusive times. These circuits are efficiently placed and routed on an FPGA to minimize total routing switches required by the architecture. Existing placement and routing algorithms are modified to efficiently place and route circuits on the architecture. Later, all unused routing switches are removed from the FPGA to generate an ASIF.

In this work a new tree-based homogeneous ASIF is presented. Exploration of tree-based ASIF is performed using a set of 16 benchmarks and experimental results have shown that tree-based ASIF is significantly smaller than an equivalent tree-based FPGA which is required to map any of these circuits. Further a comparison between mesh-based and tree-based ASIFs shows that tree-based ASIF gives better area results when compared to an equivalent mesh-based ASIF. The concept of ASIF is also extended to heterogeneous architectures where a comparison between tree-based heterogeneous ASIF with an equivalent tree-based heterogeneous FPGA is presented. Further, the comparison between mesh-based and tree-based ASIFs is also presented. The VHDL models of homogeneous and heterogeneous ASIFs are also generated using specifically developed VHDL model generator. Layout of the VHDL model is later performed using Cadence Encounter with 130 nm 6-metal layer CMOS process of ST Microelectronics.

1.3 Book Organization

The organization of this manuscript is as follows. Chapter 2 gives a detailed overview about the basic FPGA architecture and their associated design flow. Later in the chapter some current trends in the reconfigurable computing and in the FPGAs are presented also. Chapter 3 presents a detailed overview of the basic exploration environments of homogeneous mesh-based and tree-based architectures that are used in this work. This chapter also presents some new comparison results of the two architectures. Chapter 4 presents new exploration environment of tree-based heterogeneous FPGA architecture. An exploration environment for mesh-based heterogeneous FPGA architecture is also presented in this chapter. The two architectures are explored using 21 benchmarks which, based on their communication trends, are further divided into three distinct sets. Different techniques are explored using the exploration environments of two architectures and results obtained through experimentation are used for comparison between two architectures. Chapter 5 presents a new tree-based ASIF where four ASIF generation techniques are explored for a set of 16 MCNC [108] benchmarks and a comparison between tree-based ASIF and an equivalent tree-based FPGA is also presented. Later, for tree-based architecture, the effect of lookup table and arity size is explored for the most efficient technique among the four explored techniques. Further a detailed comparison between mesh-based and tree-based ASIFs is performed and finally a quality analysis of tree-based ASIF and a quality comparison between mesh-based and tree-based ASIFs is performed. Chapter 6 presents the extension of tree-based homogeneous ASIF to heterogeneous domain. Four ASIF generation techniques are explored for tree-based heterogeneous ASIF using 17 benchmarks and a comparison between tree-based ASIF and equivalent tree-based FPGA is also presented. Later experiments are performed to determine the effect of LUT and arity size on tree-based heterogeneous ASIF. After that, a comparison between mesh-based and tree-based ASIFs is performed and then a quality analysis of tree-based heterogeneous ASIF and a quality comparison between heterogeneous mesh-based and tree-based ASIFs is performed. Chapter 7 concludes this work and presents some future work.

Chapter 2
FPGA Architectures: An Overview

Field Programmable Gate Arrays (FPGAs) were first introduced almost two and a half decades ago. Since then they have seen a rapid growth and have become a popular implementation media for digital circuits. The advancement in process technology has greatly enhanced the logic capacity of FPGAs and has in turn made them a viable implementation alternative for larger and complex designs. Further, programmable nature of their logic and routing resources has a dramatic effect on the quality of final device's area, speed, and power consumption.

This chapter covers different aspects related to FPGAs. First of all an overview of the basic FPGA architecture is presented. An FPGA comprises of an array of programmable logic blocks that are connected to each other through programmable interconnect network. Programmability in FPGAs is achieved through an underlying programming technology. This chapter first briefly discusses different programming technologies. Details of basic FPGA logic blocks and different routing architectures are then described. After that, an overview of the different steps involved in FPGA design flow is given. Design flow of FPGA starts with the hardware description of the circuit which is later synthesized, technology mapped and packed using different tools. After that, the circuit is placed and routed on the architecture to complete the design flow.

The programmable logic and routing interconnect of FPGAs makes them flexible and general purpose but at the same time it makes them larger, slower and more power consuming than standard cell ASICs. However, the advancement in process technology has enabled and necessitated a number of developments in the basic FPGA architecture. These developments are aimed at further improvement in the overall efficiency of FPGAs so that the gap between FPGAs and ASICs might be reduced. These developments and some future trends are presented in the last section of this chapter.

U. Farooq et al., *Tree-Based Heterogeneous FPGA Architectures*,
DOI: 10.1007/978-1-4614-3594-5_2,

2.1 Introduction to FPGAs

Field programmable Gate Arrays (FPGAs) are pre-fabricated silicon devices that can be electrically programmed in the field to become almost any kind of digital circuit or system. For low to medium volume productions, FPGAs provide cheaper solution and faster time to market as compared to Application Specific Integrated Circuits (ASIC) which normally require a lot of resources in terms of time and money to obtain first device. FPGAs on the other hand take less than a minute to configure and they cost anywhere around a few hundred dollars to a few thousand dollars. Also for varying requirements, a portion of FPGA can be partially reconfigured while the rest of an FPGA is still running. Any future updates in the final product can be easily upgraded by simply downloading a new application bitstream. However, the main advantage of FPGAs i.e. flexibility is also the major cause of its draw back. Flexible nature of FPGAs makes them significantly larger, slower, and more power consuming than their ASIC counterparts. These disadvantages arise largely because of the programmable routing interconnect of FPGAs which comprises of almost 90% of total area of FPGAs. But despite these disadvantages, FPGAs present a compelling alternative for digital system implementation due to their less time to market and low volume cost.

Normally FPGAs comprise of:

- Programmable logic blocks which implement logic functions.
- Programmable routing that connects these logic functions.
- I/O blocks that are connected to logic blocks through routing interconnect and that make off-chip connections.

A generalized example of an FPGA is shown in Fig. 2.1 where configurable logic blocks (CLBs) are arranged in a two dimensional grid and are interconnected by programmable routing resources. I/O blocks are arranged at the periphery of the grid and they are also connected to the programmable routing interconnect. The "programmable/reconfigurable" term in FPGAs indicates their ability to implement a new function on the chip after its fabrication is complete. The reconfigurability/programmability of an FPGA is based on an underlying programming technology, which can cause a change in behavior of a pre-fabricated chip after its fabrication.

2.2 Programming Technologies

There are a number of programming technologies that have been used for reconfigurable architectures. Each of these technologies have different characteristics which in turn have significant effect on the programmable architecture. Some of the well known technologies include static memory [122], flash [54], and anti-fuse [61].

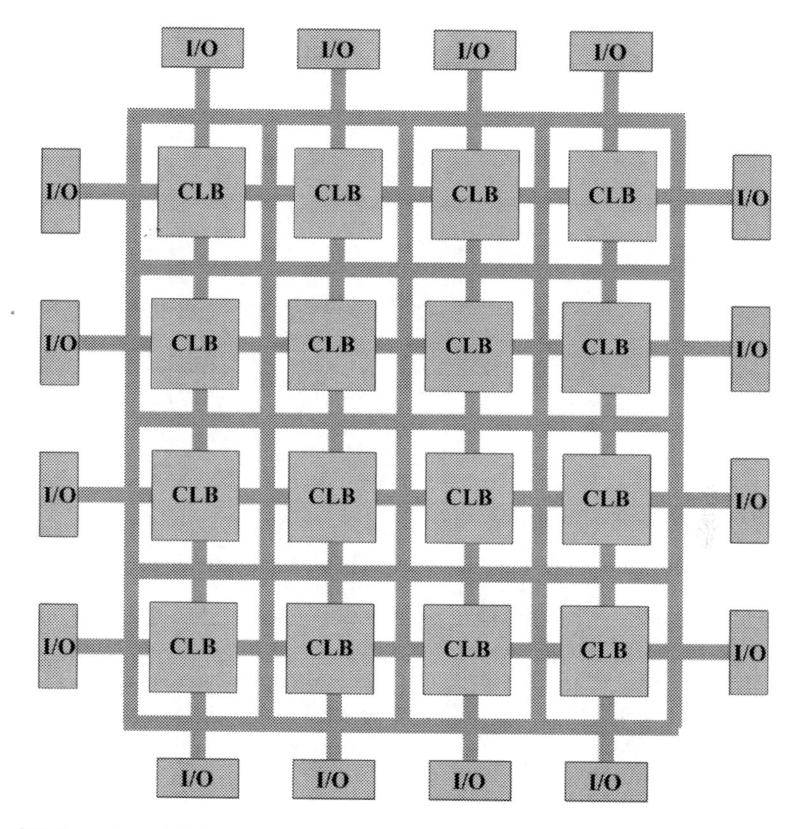

Fig. 2.1 Overview of FPGA architecture [22]

2.2.1 SRAM-Based Programming Technology

Static memory cells are the basic cells used for SRAM-based FPGAs. Most commercial vendors [76, 126] use static memory (SRAM) based programming technology in their devices. These devices use static memory cells which are divided throughout the FPGA to provide configurability. An example of such memory cell is shown in Fig. 2.2. In an SRAM-based FPGA, SRAM cells are mainly used for following purposes:

1. To program the routing interconnect of FPGAs which are generally steered by small multiplexors.
2. To program Configurable Logic Blocks (CLBs) that are used to implement logic functions.

SRAM-based programming technology has become the dominant approach for FPGAs because of its re-programmability and the use of standard CMOS process technology and therefore leading to increased integration, higher speed and lower

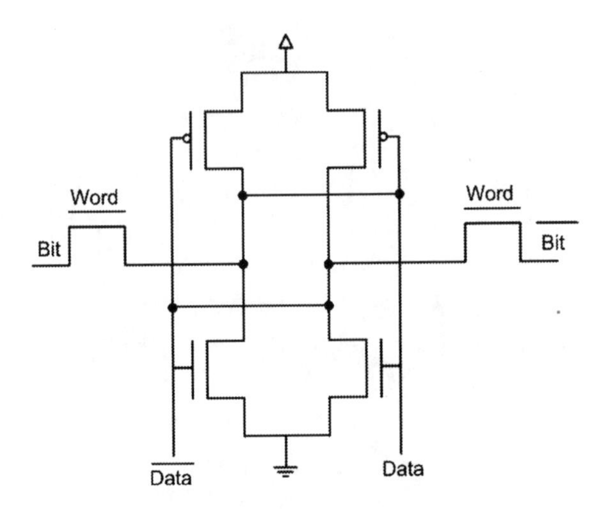

Fig. 2.2 Static memory cell

dynamic power consumption of new process with smaller geometry. There are however a number of drawbacks associated with SRAM-based programming technology. For example an SRAM cell requires 6 transistors which makes the use of this technology costly in terms of area compared to other programming technologies. Further SRAM cells are volatile in nature and external devices are required to permanently store the configuration data. These external devices add to the cost and area overhead of SRAM-based FPGAs.

2.2.2 Flash Programming Technology

One alternative to the SRAM-based programming technology is the use of flash or EEPROM based programming technology. Flash-based programming technology offers several advantages. For example, this programming technology is non-volatile in nature. Flash-based programming technology is also more area efficient than SRAM-based programming technology. Flash-based programming technology has its own disadvantages also. Unlike SRAM-based programming technology, flash-based devices can not be reconfigured/reprogrammed an infinite number of times. Also, flash-based technology uses non-standard CMOS process.

2.2.3 Anti-fuse Programming Technology

An alternative to SRAM and flash-based technologies is anti-fuse programming technology. The primary advantage of anti-fuse programming technology is its low area. Also this technology has lower on resistance and parasitic capacitance than other two programming technologies. Further, this technology is non-volatile in nature. There are however significant disadvantages associated with this programming technology. For example, this technology does not make use of standard CMOS process. Also, anti-fuse programming technology based devices can not be reprogrammed.

In this section, an overview of three commonly used programming technologies is given where all of them have their advantages and disadvantages. Ideally, one would like to have a programming technology which is reprogrammable, non-volatile, and that uses a standard CMOS process. Apparently, none of the above presented technologies satisfy these conditions. However, SRAM-based programming technology is the most widely used programming technology. The main reason is its use of standard CMOS process and for this very reason, it is expected that this technology will continue to dominate the other two programming technologies.

2.3 Configurable Logic Block

A configurable logic block (CLB) is a basic component of an FPGA that provides the basic logic and storage functionality for a target application design. In order to provide the basic logic and storage capability, the basic component can be either a transistor or an entire processor. However, these are the two extremes where at one end the basic component is very fine-grained (in case of transistors) and requires large amount of programmable interconnect which eventually results in an FPGA that suffers from area-inefficiency, low performance and high power consumption. On the other end (in case of processor), the basic logic block is very coarse-grained and can not be used to implement small functions as it will lead to wastage of resources. In between these two extremes, there exists a spectrum of basic logic blocks. Some of them include logic blocks that are made of NAND gates [101], an interconnection of multiplexors [44], lookup table (LUT) [121] and PAL style wide input gates [124]. Commercial vendors like Xilinx and Altera use LUT-based CLBs to provide basic logic and storage functionality. LUT-based CLBs provide a good trade-off between too fine-grained and too coarse-grained logic blocks. A CLB can comprise of a single basic logic element (BLE), or a cluster of locally interconnected BLEs (as shown in Fig. 2.4). A simple BLE consists of a LUT, and a Flip-Flop. A LUT with k inputs (LUT-k) contains 2^k configuration bits and it can implement any k-input boolean function. Figure 2.3 shows a simple BLE comprising of a 4 input LUT (LUT-4) and a D-type Flip-Flop. The LUT-4 uses 16 SRAM bits to implement any 4 inputs boolean function. The output of LUT-4 is connected to an optional Flip-Flop. A multiplexor selects the BLE output to be either the output of a Flip-Flop or the LUT-4.

A CLB can contain a cluster of BLEs connected through a local routing network. Figure 2.4 shows a cluster of 4 BLEs; each BLE contains a LUT-4 and a Flip-Flop. The BLE output is accessible to other BLEs of the same cluster through a local routing network. The number of output pins of a cluster are equal to the total number of BLEs in a cluster (with each BLE having a single output). However, the number of input pins of a cluster can be less than or equal to the sum of input pins required

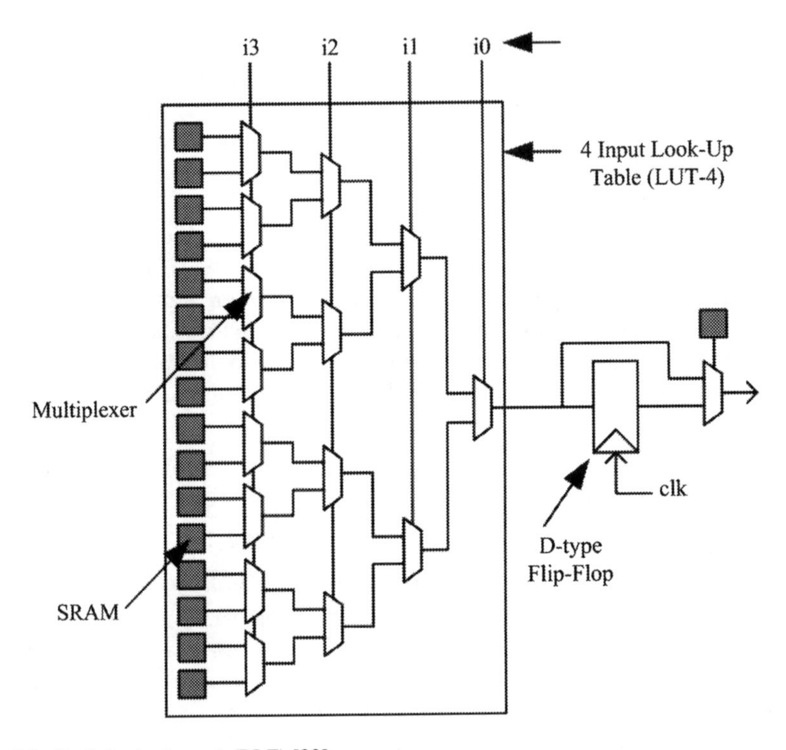

Fig. 2.3 Basic logic element (BLE) [22]

by all the BLEs in the cluster. Modern FPGAs contain typically 4 to 10 BLEs in a single cluster. Although here we have discussed only basic logic blocks, many modern FPGAs contain a heterogeneous mixture of blocks, some of which can only be used for specific purposes. Theses specific purpose blocks, also referred here as hard blocks, include memory, multipliers, adders and DSP blocks etc. Hard blocks are very efficient at implementing specific functions as they are designed optimally to perform these functions, yet they end up wasting huge amount of logic and routing resources if unused. A detailed discussion on the use of heterogeneous mixture of blocks for implementing digital circuits is presented in Chap. 4 where both advantages and disadvantages of heterogeneous FPGA architectures and a remedy to counter the resource loss problem are discussed in detail.

2.4 FPGA Routing Architectures

As discussed earlier, in an FPGA, the computing functionality is provided by its programmable logic blocks and these blocks connect to each other through programmable routing network. This programmable routing network provides routing

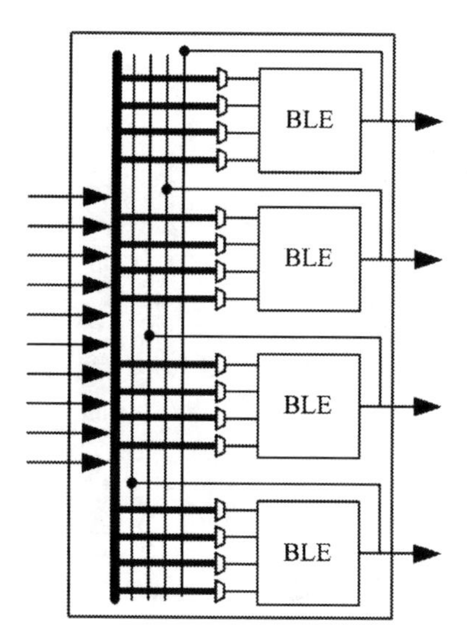

Fig. 2.4 A configurable logic block (CLB) having four BLEs [22]

connections among logic blocks and I/O blocks to implement any user-defined circuit. The routing interconnect of an FPGA consists of wires and programmable switches that form the required connection. These programmable switches are configured using the programmable technology.

Since FPGA architectures claim to be potential candidate for the implementation of any digital circuit, their routing interconnect must be very flexible so that they can accommodate a wide variety of circuits with widely varying routing demands. Although the routing requirements vary from circuit to circuit, certain common characteristics of these circuits can be used to optimally design the routing interconnect of FPGA architecture. For example most of the designs exhibit locality, hence requiring abundant short wires. But at the same time there are some distant connections, which leads to the need for sparse long wires. So, care needs to be taken into account while designing routing interconnect for FPGA architectures where we have to address both flexibility and efficiency. The arrangement of routing resources, relative to the arrangement of logic blocks of the architecture, plays a very important role in the overall efficiency of the architecture. This arrangement is termed here as global routing architecture whereas the microscopic details regarding the switching topology of different switch blocks is termed as detailed routing architecture. On the basis of the global arrangement of routing resources of the architecture, FPGA architectures can be categorized as either hierarchical [4] or island-style [22]. In this section, we present a detailed overview of both routing architectures.

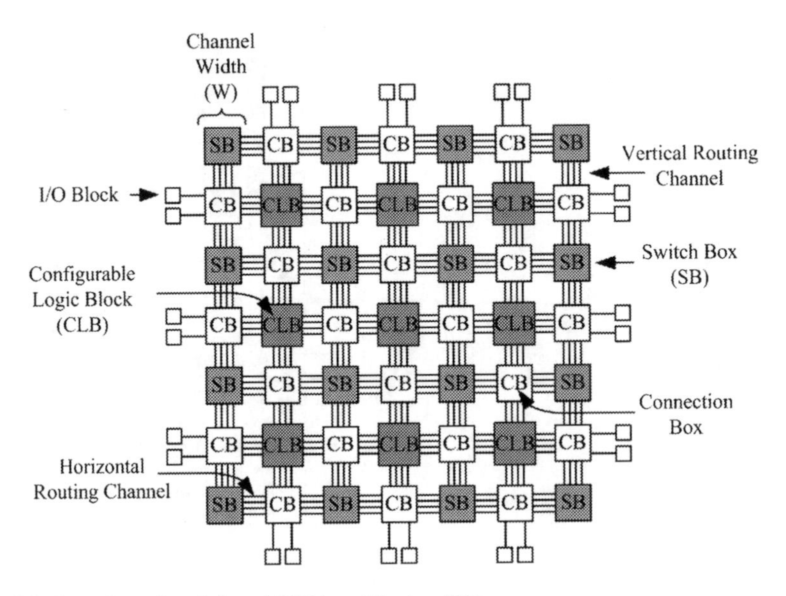

Fig. 2.5 Overview of mesh-based FPGA architecture [22]

2.4.1 Island-Style Routing Architecture

Figure 2.5 shows a traditional island-style FPGA architecture (also termed as mesh-based FPGA architecture). This is the most commonly used architecture among academic and commercial FPGAs. It is called island-style architecture because in this architecture configurable logic blocks look like islands in a sea of routing interconnect. In this architecture, configurable logic blocks (CLBs) are arranged on a 2D grid and are interconnected by a programmable routing network. The Input/Output (I/O) blocks on the periphery of FPGA chip are also connected to the programmable routing network. The routing network comprises of pre-fabricated wiring segments and programmable switches that are organized in horizontal and vertical routing channels.

The routing network of an FPGA occupies 80–90% of total area, whereas the logic area occupies only 10–20% area [22]. The flexibility of an FPGA is mainly dependent on its programmable routing network. A mesh-based FPGA routing network consists of horizontal and vertical routing tracks which are interconnected through switch boxes (SB). Logic blocks are connected to the routing network through connection boxes (CB). The flexibility of a connection box (Fc) is the number of routing tracks of adjacent channel which are connected to the pin of a block. The connectivity of input pins of logic blocks with the adjacent routing channel is called as Fc(in); the connectivity of output pins of the logic blocks with the adjacent routing channel is called as Fc(out). An Fc(in) equal to 1.0 means that all the tracks of adjacent routing channel are connected to the input pin of the logic block. The flexibility of switch box (Fs) is the total number of tracks with which every track entering in the switch

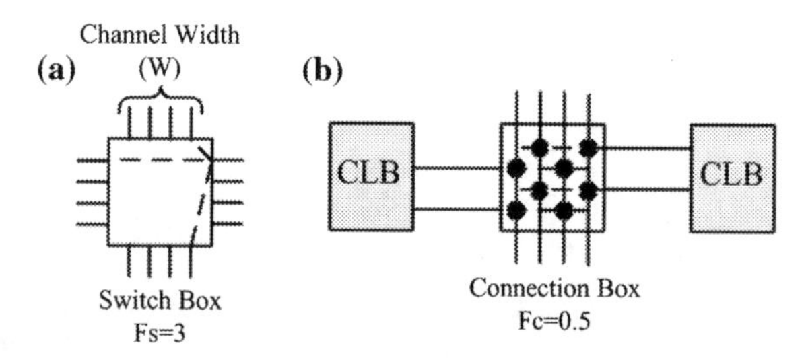

Fig. 2.6 Example of switch and connection box

box connects to. The number of tracks in routing channel is called the channel width of the architecture. Same channel width is used for all horizontal and vertical routing channels of the architecture. An example explaining the switch box, connection box flexibilities, and routing channel width is shown in Fig. 2.6. In this figure switch box has Fs = 3 as each track incident on it is connected to 3 tracks of adjacent routing channels. Similarly, connection box has Fc(in) = 0.5 as each input of the logic block is connected to 50% of the tracks of adjacent routing channel.

The routing tracks connected through a switch box can be bidirectional or unidirectional (also called as directional) tracks. Figure 2.7 shows a bidirectional and a unidirectional switch box having Fs equal to 3. The input tracks (or wires) in both these switch boxes connect to 3 other tracks of the same switch box. The only limitation of unidirectional switch box is that their routing channel width must be in multiples of 2.

Generally, the output pins of a block can connect to any routing track through pass transistors. Each pass transistor forms a tristate output that can be independently turned on or off. However, single-driver wiring technique can also be used to connect output pins of a block to the adjacent routing tracks. For single-driver wiring, tristate elements cannot be used; the output of block needs to be connected to the neighboring routing network through multiplexors in the switch box. Modern commercial FPGA architectures have moved towards using single-driver, directional routing tracks. Authors in [51] show that if single-driver directional wiring is used instead of bidirectional wiring, 25% improvement in area, 9% in delay and 32% in area-delay can be achieved. All these advantages are achieved without making any major changes in the FPGA CAD flow.

In mesh-based FPGAs, multi-length wires are created to reduce delay. Figure 2.8 shows an example of different length wires. Longer wire segments span multiple blocks and require fewer switches, thereby reducing routing area and delay. However, they also decrease routing flexibility, which reduces the probability to route a hardware circuit successfully. Modern commercial FPGAs commonly use a combination of long and short wires to balance flexibility, area and delay of the routing network.

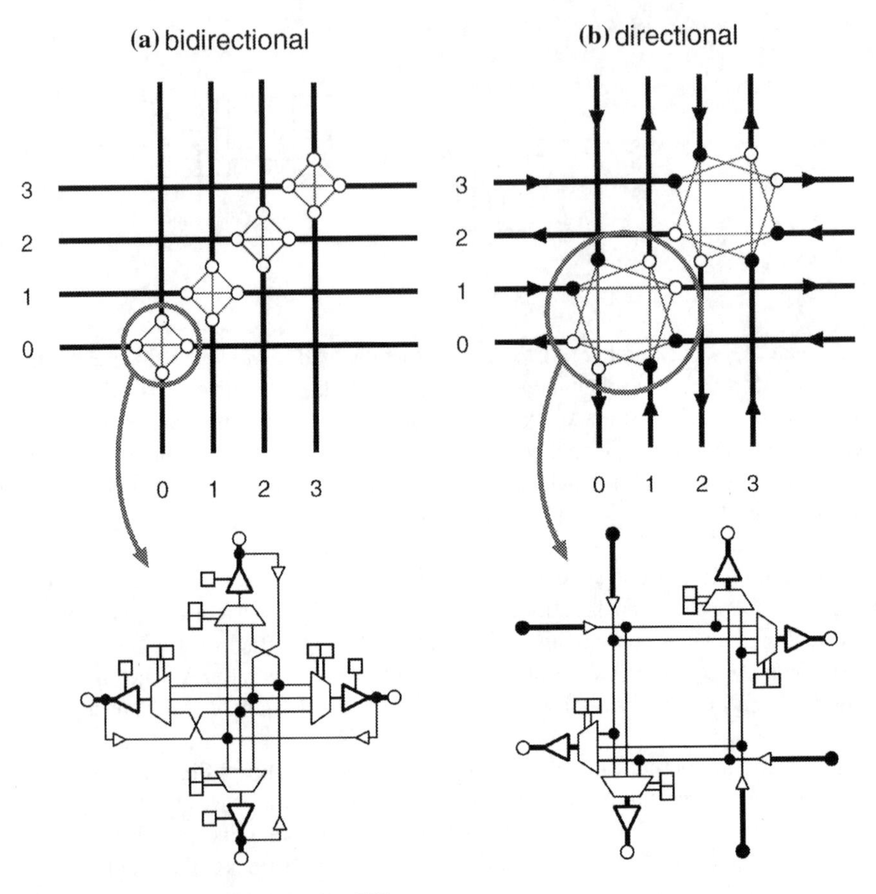

Fig. 2.7 Switch block, length 1 wires [51]

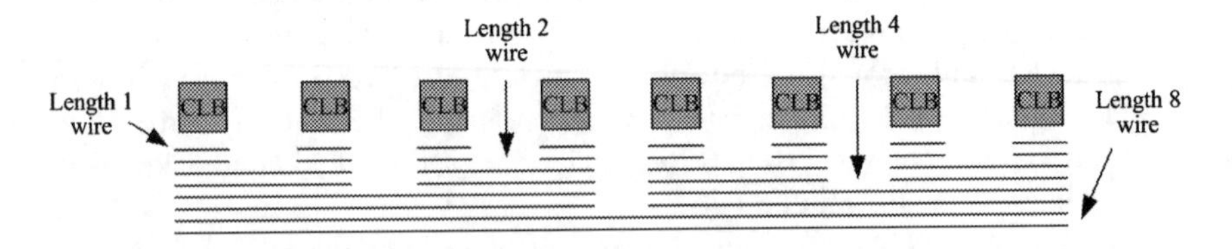

Fig. 2.8 Channel segment distribution

2.4.1.1 Altera's Stratix II Architecture

Until now, we have presented a general overview about island-style routing architecture. Now we present a commercial example of this kind of architectures. Altera's Stratix II [106] architecture is an industrial example of an island-style FPGA (Fig. 2.9). The logic structure consists of LABs (Logic Array Blocks), memory blocks, and digital signal processing (DSP) blocks. LABs are used to

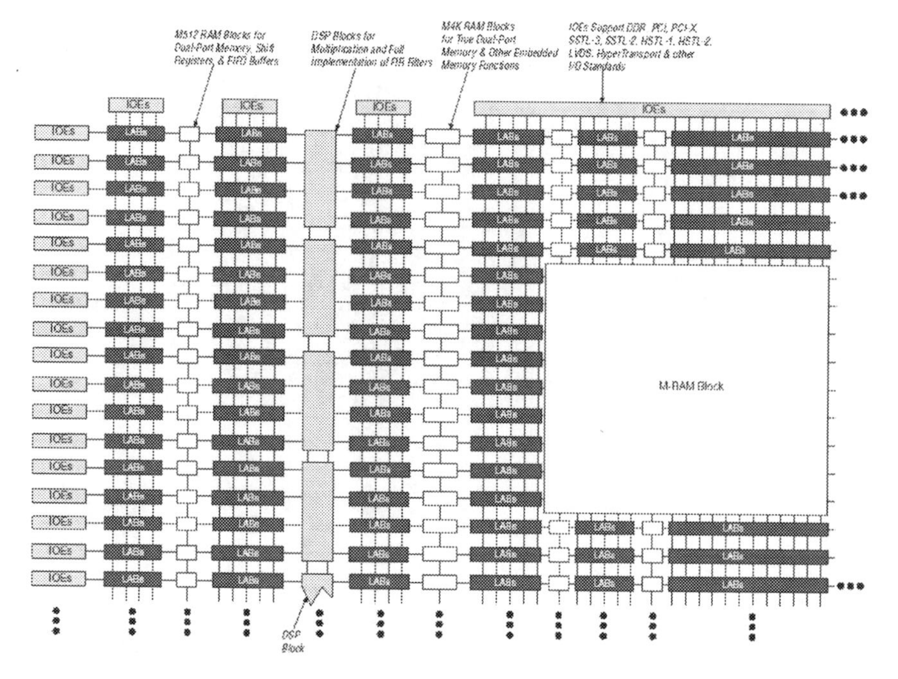

Fig. 2.9 Altera's stratix-II block diagram

implement general-purpose logic, and are symmetrically distributed in rows and columns throughout the device fabric. The DSP blocks are custom designed to implement full-precision multipliers of different granularities, and are grouped into columns. Input- and output-only elements (IOEs) represent the external interface of the device. IOEs are located along the periphery of the device.

Each Stratix II LAB consists of eight Adaptive Logic Modules (ALMs). An ALM consists of 2 adaptive LUTs (ALUTs) with eight inputs altogether. Construction of an ALM allows implementation of 2 separate 4-input Boolean functions. Further, an ALM can also be used to implement any six-input Boolean function, and some seven-input functions. In addition to lookup tables, an ALM provides 2 programmable registers, 2 dedicated full-adders, a carry chain, and a register-chain. Full-adders and carry chain can be used to implement arithmetic operations, and the register-chain is used to build shift registers. Outputs of an ALM drive all types of interconnect provided by the Stratix II device. Figure 2.10 illustrates a LAB interconnect interface.

Interconnections between LABs, RAM blocks, DSP blocks and the IOEs are established using the Multi-track interconnect structure. This interconnect structure consists of wire segments of different lengths and speeds. The interconnect wire-segments span fixed distances, and run in the horizontal (row interconnects) and vertical (column interconnects) directions. The row interconnects (Fig. 2.11) can be used to route signals between LABs, DSP blocks, and memory blocks in the same row. Row interconnect resources are of the following types:

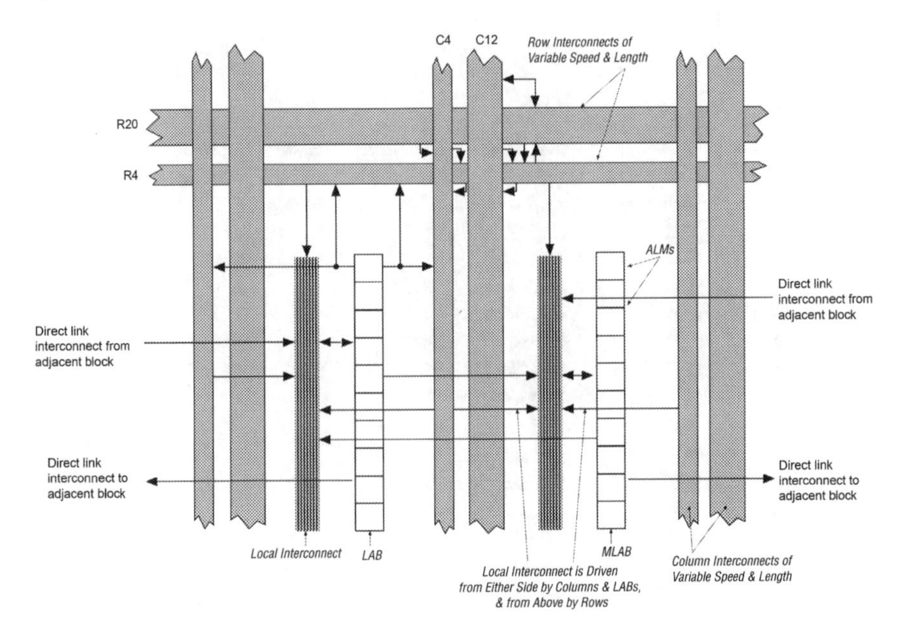

Fig. 2.10 Stratix-II logic array block (LAB) structure

- Direct connections between LABs and adjacent blocks.
- R4 resources that span 4 blocks to the left or right.
- R24 resources that provide high-speed access across 24 columns.

Each LAB owns its set of R4 interconnects. A LAB has approximately equal numbers of driven-left and driven-right R4 interconnects. An R4 interconnect that is driven to the left can be driven by either the primary LAB (Fig. 2.11) or the adjacent LAB to the left.

Similarly, a driven-right R4 interconnect may be driven by the primary LAB or the LAB immediately to its right. Multiple R4 resources can be connected to each other to establish longer connections within the same row. R4 interconnects can also drive C4 and C16 column interconnects, and R24 high speed row resources.

Column interconnect structure is similar to row interconnect structure. Column interconnects include:

- Carry chain interconnects within a LAB, and from LAB to LAB in the same column.
- Register chain interconnects.
- C4 resources that span 4 blocks in the up and down directions.
- C16 resources for high-speed vertical routing across 16 rows.

Carry chain and register chain interconnects are separated from local interconnect (Fig. 2.10) in a LAB. Each LAB has its own set of driven-up and driven-down C4 interconnects. C4 interconnects can also be driven by the LABs that are immediately

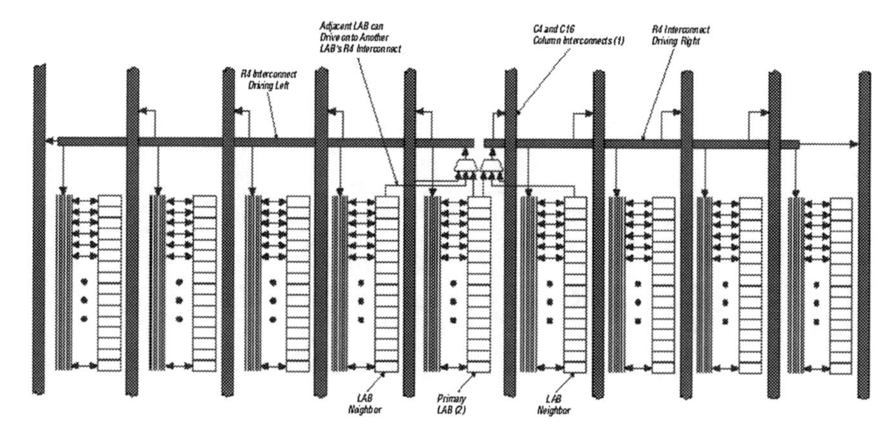

Fig. 2.11 R4 interconnect connections

adjacent to the primary LAB. Multiple C4 resources can be connected to each other to form longer connections within a column, and C4 interconnects can also drive row interconnects to establish column-to-column interconnections. C16 interconnects are high-speed vertical resources that span 16 LABs. A C16 interconnect can drive row and column interconnects at every fourth LAB. A LAB local interconnect structure cannot be directly driven by a C16 interconnect; only C4 and R4 interconnects can drive a LAB local interconnect structure. Figure 2.12 shows the C4 interconnect structure in the Stratix II device.

2.4.2 Hierarchical Routing Architecture

Most logic designs exhibit locality of connections; hence implying a hierarchy in placement and routing of connections between different logic blocks. Hierarchical routing architectures exploit this locality by dividing FPGA logic blocks into separate groups/clusters. These clusters are recursively connected to form a hierarchical structure. In a hierarchical architecture (also termed as tree-based architecture), connections between logic blocks within same cluster are made by wire segments at the lowest level of hierarchy. However, the connection between blocks residing in different groups require the traversal of one or more levels of hierarchy. In a hierarchical architecture, the signal bandwidth varies as we move away from the bottom level and generally it is widest at the top level of hierarchy. The hierarchical routing architecture has been used in a number of commercial FPGA families including Altera Flex10K [10], Apex [15] and ApexII [16] architectures. We assume that Multilevel hierarchical interconnect regroups architectures with more than 2 levels of hierarchy and Tree-based ones.

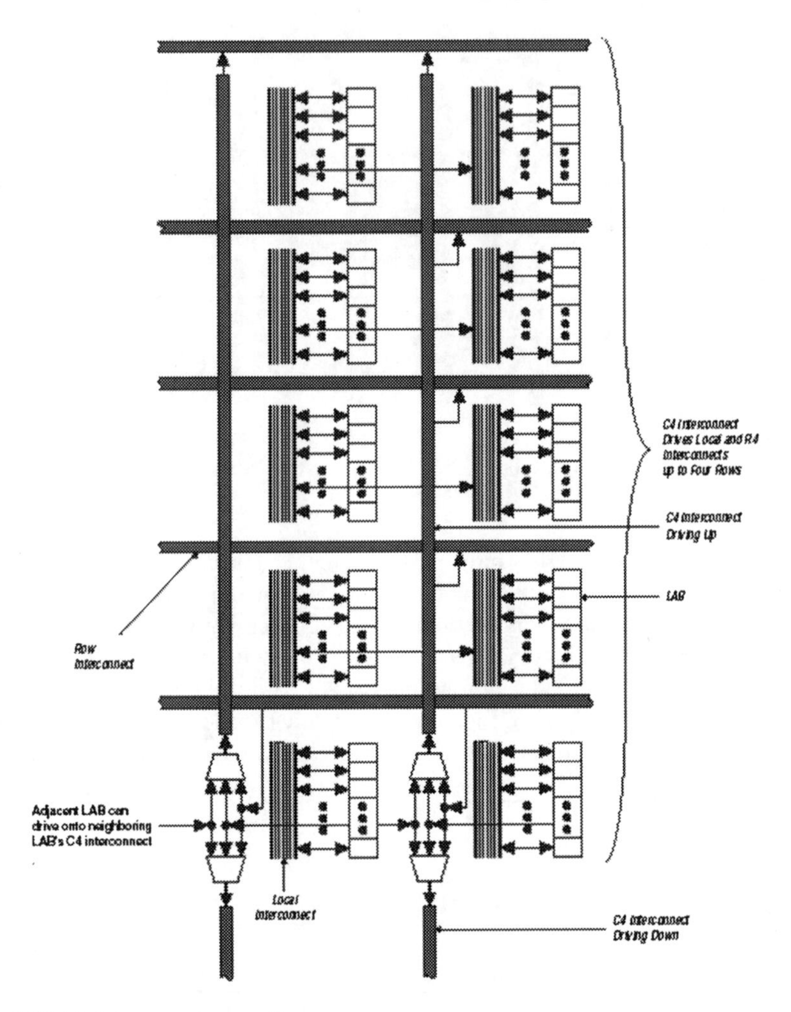

Fig. 2.12 C4 interconnect connections

2.4.2.1 HFPGA: Hierarchical FPGA

In the hierarchical FPGA called HFPGA, LBs are grouped into clusters. Clusters are then grouped recursively together (see Fig. 2.13). The clustered VPR mesh architecture [22] has a Hierarchical topology with only two levels. Here we consider multilevel hierarchical architectures with more than 2 levels. In [1] and [129] various hierarchical structures were discussed. The HFPGA routability depends on switch boxes topologies. HFPGAs comprising fully populated switch boxes ensure 100% routability but are very penalizing in terms of area. In [129] authors explored the HFPGA architecture, investigating how the switch pattern can be partly depopulated while maintaining a good routability.

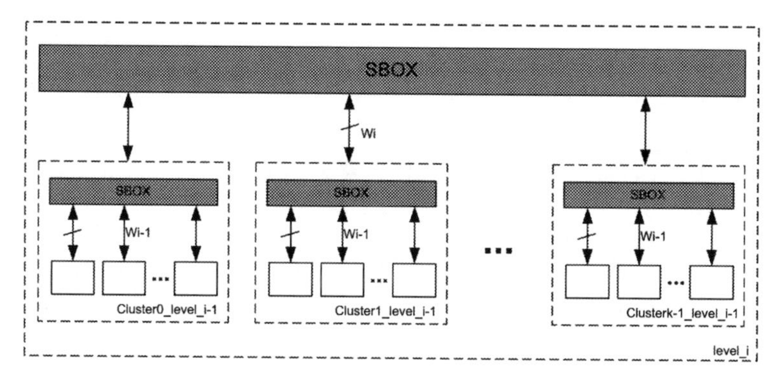

Fig. 2.13 Hierarchical FPGA topology

2.4.2.2 HSRA: Hierarchical Synchronous Reconfigurable Array

An example of an academic hierarchical routing architecture is shown in Fig. 2.14. It has a strictly hierarchical, tree-based interconnect structure. In this architecture, the only wire segments that directly connect to the logic units are located at the leaves of the interconnect tree. All other wire segments are decoupled from the logic structure. A logic block of this architecture consists of a pair of 2-input Look Up Table (2-LUT) and a D-type Flip Flop (D-FF). The input-pin connectivity is based on a choose-k strategy [4], and the output pins are fully connected. The richness of this interconnect structure is defined by its base channel width c and interconnect growth rate p. The base channel width c is defined as the number of tracks at the leaves of the interconnect Tree (in Fig. 2.14, $c = 3$). Growth rate p is defined as the rate at which the interconnect bandwidth grows towards the upper levels. The interconnect growth rate can be realized either using non-compressing or compressing switch blocks. The details regarding these switch blocks is as follows:

- Non-compressing (2:1) switch blocks—The number of tracks at the upper level are equal to the sum of the number of tracks of the children at lower level. For example, in Fig. 2.14, non-compressing switch blocks are used between levels 1, 2 and levels 3, 4.
- Compressing (1:1) switch blocks—The number of tracks at the upper level are equal to the number of tracks of either child at the lower level. For example, in Fig. 2.14, compressing switch blocks are used between levels 2 and 3.

A repeating combination of non-compressing and compressing switch blocks can be used to realize any value of p less than one. For example, a repeating pattern of (2:1, 1:1) switch blocks realizes $p = 0.5$, while the pattern (2:1, 2:1, 1:1) realizes $p = 0.67$. An architecture that has only 2:1 switch blocks provides a growth rate of $p = 1$.

Another hierarchical routing architecture is presented in [132] where the global routing architecture (i.e. the position of routing resources relative to logic resources

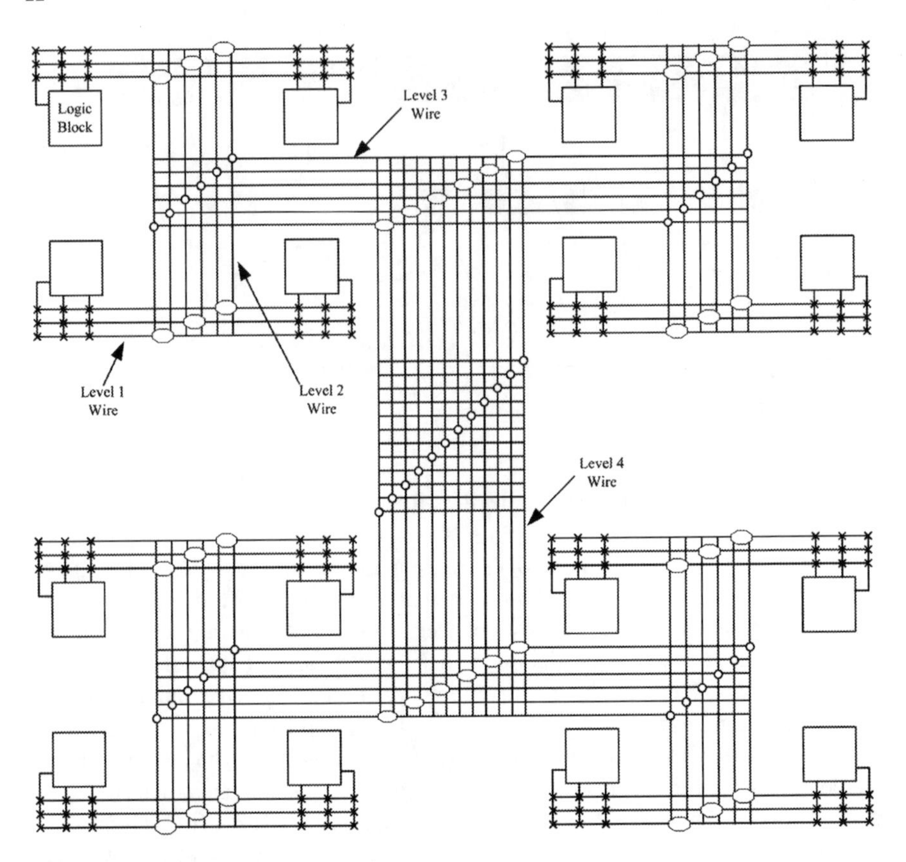

Fig. 2.14 Example of hierarchical routing architecture [4]

of the architecture) remains the same as in [4]. However, there are several key differences at the level of detailed routing architecture (i.e. the way the routing resources are connected to each other, flexibility of switch blocks etc.) that separate the two architectures. For example the architecture shown in Fig. 2.14 has one bidirectional interconnect that uses bidirectional switches and it supports only arity-2 (i.e. each cluster can contain only two sub-clusters). On contrary, the architecture presented in [132] supports two separate unidirectional interconnect networks: one is downward interconnect whereas other is upward interconnect network. Further this architecture is more flexible as it can support logic blocks with different sizes and also the clusters/groups of the routing architecture can have different arity sizes. Further details of this architecture, from now on alternatively termed as tree-based architecture, are presented in next chapter.

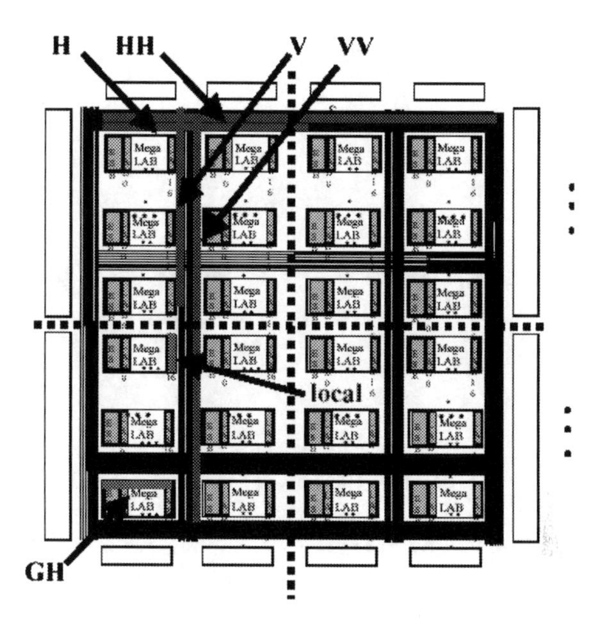

Fig. 2.15 The APEX programmable logic Devices [87]

2.4.2.3 APEX: Altera

APEX architecture is a commercial product from Altera Corporation which includes 3 levels of interconnect hierarchy. Figure 2.15 shows a diagram of the APEX 20K400 programmable logic device. The basic logic-element (LE) is a 4-input LUT and DFF pair. Groups of 10 LEs are grouped into a logic-array-block or LAB. Interconnect within a LAB is complete, meaning that a connection from the output of any LE to the input of another LE in its LAB always exists, and any signal entering the input region can reach every LE.

Groups of 16 LABs form a MegaLab. Interconnect within a MegaLab requires an LE to drive a GH (MegaLab global H) line, a horizontal line, which switches into the input region of any other LAB in the same MegaLab. Adjacent LABs have the ability to interleave their input regions, so an LE in LAB_i can usually drive LAB_{i+1} without using a GH line. A 20K400 MegaLab contains 279 GH lines.

The top-level architecture is a 4 by 26 array of MegaLabs. Communication between MegaLabs is accomplished by global H (horizontal) and V (vertical) wires, that switch at their intersection points. The H and V lines are segmented by a bidirectional segmentation buffer at the horizontal and vertical centers of the chip. In Fig. 2.15, We denote the use of a single (half-chip) line as H or V and a double or full-chip line through the segmentation buffer as HH or VV. The 20K400 contains 100 H lines per MegaLab row, and 80 V lines per LAB-column.

In this section, so far we have given an overview of the two routing architectures that are commonly employed in FPGAs. Both architectures have their positive and negative points. For example, hierarchical routing architectures exploit the

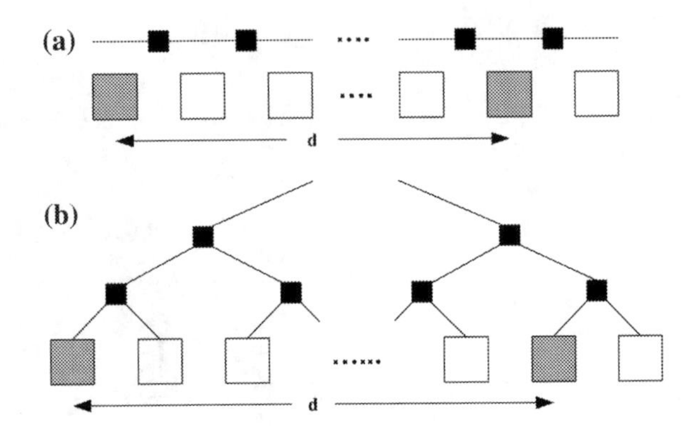

Fig. 2.16 **a** Number of series switches in a mesh structure **b** Number of series switches in a tree structure

locality exhibited by the most of the designs and in turn offer smaller delays and more predictable routing compared to island-style architectures. The speed of a net is determined by the number of routing switches it has to pass and the length of wires. In a mesh-based architecture, the number of segments increase linearly with manhattan distance d between the logic blocks to be connected. However, for tree-based architecture the distance d between the blocks to be connected increases in a logarithmic manner [82]. This fact is illustrated in Fig. 2.16. On the other hand, scalability is an issue in hierarchical routing architectures and there might be some design mapping issues. But in the case of mesh-based architecture, there are no such issues as it offers a tile-based layout where a tile once formed can be replicated horizontally and vertically to make as large architecture as we wish.

2.5 Software Flow

FPGA architectures have been intensely investigated over the past two decades. A major aspect of FPGA architecture research is the development of Computer Aided Design (CAD) tools for mapping applications to FPGAs. It is well established that the quality of an FPGA-based implementation is largely determined by the effectiveness of accompanying suite of CAD tools. Benefits of an otherwise well designed, feature rich FPGA architecture might be impaired if the CAD tools cannot take advantage of the features that the FPGA provides. Thus, CAD algorithm research is essential to the necessary architectural advancement to narrow the performance gaps between FPGAs and other computational devices like ASICs.

The software flow (CAD flow) takes an application design description in a Hardware Description Language (HDL) and converts it to a stream of bits that is eventually programmed on the FPGA. The process of converting a circuit description into a format that can be loaded into an FPGA can be roughly divided into five distinct steps, namely: synthesis, technology mapping, mapping, placement and routing. The final output of FPGA CAD tools is a bitstream that configures the state of the memory

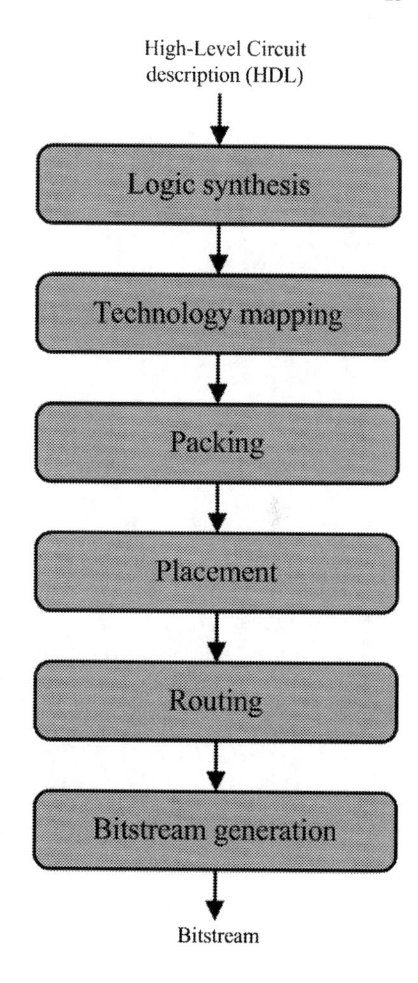

Fig. 2.17 FPGA software flow

bits in an FPGA. The state of these bits determines the logical function that the FPGA implements. Figure 2.17 shows a generalized software flow for programming an application circuit on an FPGA architecture. A description of various modules of software flow is given in the following part of this section. The details of these modules are generally indifferent to the kind of routing architecture being used and they are applicable to both architectures described earlier unless otherwise specified.

2.5.1 Logic Synthesis

The flow of FPGA starts with the logic synthesis of the netlist being mapped on it. Logic synthesis [26, 27] transforms an HDL description (VHDL or Verilog) into a set of boolean gates and Flip-Flops. The synthesis tools transform the

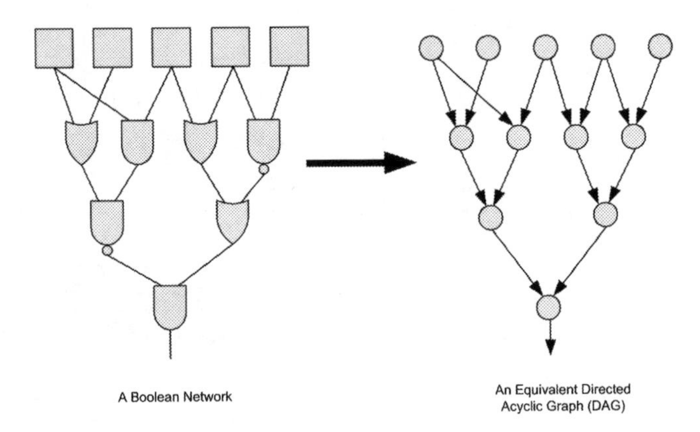

Fig. 2.18 Directed acyclic graph representation of a circuit

register-transfer-level (RTL) description of a design into a hierarchical boolean network. Various technology-independent techniques are applied to optimize the boolean network. The typical cost function of technology-independent optimizations is the total literal count of the factored representation of the logic function. The literal count correlates very well with the circuit area. Further details of logic synthesis are beyond the scope of this book.

2.5.2 Technology Mapping

The output from synthesis tools is a circuit description of Boolean logic gates, flip-flops and wiring connections between these elements. The circuit can also be represented by a Directed Acyclic Graph (DAG). Each node in the graph represents a gate, flip-flop, primary input or primary output. Each edge in the graph represents a connection between two circuit elements. Figure 2.18 shows an example of a DAG representation of a circuit. Given a library of cells, the technology mapping problem can be expressed as finding a network of cells that implements the Boolean network. In the FPGA technology mapping problem, the library of cells is composed of k-input LUTs and flip-flops. Therefore, FPGA technology mapping involves transforming the Boolean network into k-bounded cells. Each cell can then be implemented as an independent k-LUT. Figure 2.19 shows an example of transforming a Boolean network into k-bounded cells. Technology mapping algorithms can optimize a design for a set of objectives including depth, area or power. The FlowMap algorithm [64] is the most widely used academic tool for FPGA technology mapping. FlowMap is a breakthrough in FPGA technology mapping because it is able to find a depth-optimal solution in polynomial time. FlowMap guarantees depth optimality at the expense of logic duplication. Since the introduction of FlowMap, numerous technology mappers have been designed that optimize for area and run-time while still maintaining

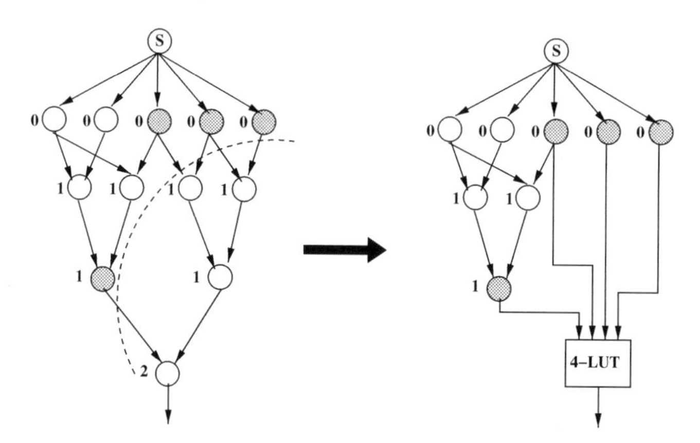

Fig. 2.19 Example of technology mapping

the depth-optimality of the circuit [65–67]. The result of the technology mapping step generates a network of k-bounded LUTs and flip-flops.

2.5.3 *Clustering/Packing*

The logic elements in a Mesh-based FPGA are typically arranged in two levels of hierarchy. The first level consists of logic blocks (LBs) which are k-input LUT and flip-flop pairs. The second level hierarchy groups k LBs together to form logic blocks clusters. The clustering phase of the FPGA CAD flow is the process of forming groups of k LBs. These clusters can then be mapped directly to a logic element on an FPGA. Figure 2.20 shows an example of the clustering process.

Clustering algorithms can be broadly categorized into three general approaches, namely top-down [39, 78], depth-optimal [84, 100] and bottom-up [14, 17, 43]. Top-down approaches partition the LBs into clusters by successively subdividing the network or by iteratively moving LBs between parts. Depth-optimal solutions attempt to minimize delay at the expense of logic duplication. Bottom-up approaches are generally preferred for FPGA CAD tools due to their fast run times and reasonable timing delays. They only consider local connectivity information and can easily satisfy clusters pin constraints. Top-down approaches offer the best solutions; however, their computational complexity can be prohibitive.

2.5.3.1 Bottom-up Approaches

Bottom-up approaches build clusters sequentially one at a time. The process starts by choosing an LB which acts as a cluster seed. LBs are then greedily selected and added to the cluster, applying various attraction functions. The VPack [14] attraction

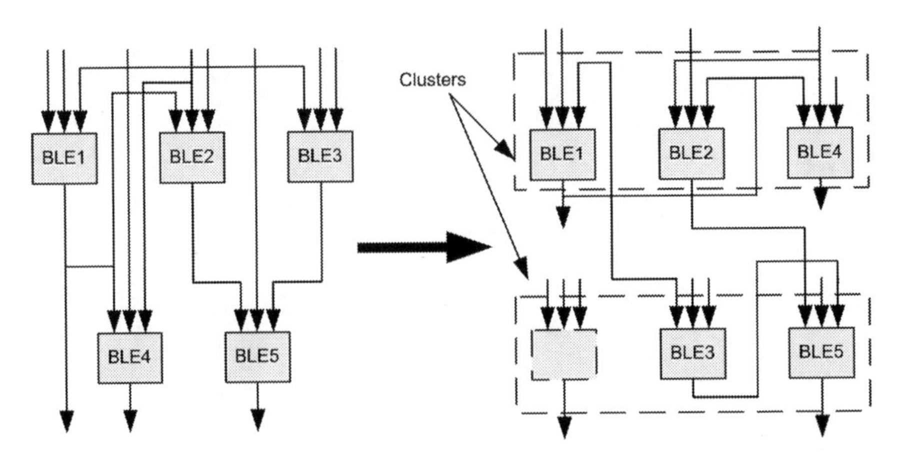

Fig. 2.20 Example of packing

function is based on the number of shared nets between a candidate LB and the LBs that are already in the cluster. For each cluster, the attraction function is used to select a seed LB from the set of all LBs that have not already been packed. After packing a seed LB into the new cluster, a second attraction function selects new LBs to pack into the cluster. LBs are packed into the cluster until the cluster reaches full capacity or all cluster inputs have been used. If all cluster inputs become occupied before this cluster reaches full capacity, a hill-climbing technique is applied, searching for LBs that do not increase the number of inputs used by the cluster. The VPack pseudo-code is outlined in algorithm 2.1.

T-VPack [22] is a timing-driven version of VPack which gives added weight to grouping LBs on the critical path together. The algorithm is identical to VPack, however, the attraction functions which select the LBs to be packed into the clusters are different. The VPack seed function chooses LBs with the most used inputs, whereas the T-VPack seed function chooses LBs that are on the most critical path. VPack's second attraction function chooses LBs with the largest number of connections with the LBs already packed into the cluster. T-VPack's second attraction function has two components for a LB B being considered for cluster C:

$$Attraction(B, C) = \alpha.Crit(B) + (1 - \alpha)\frac{\mid Nets(B) \cap Nets(C) \mid}{G} \qquad (2.1)$$

where $Crit(B)$ is a measure of how close LB B is to being on the critical path, $Nets(B)$ is the set of nets connected to LB B, $Nets(C)$ is the set of nets connected to the LBs already selected for cluster C, α is a user-defined constant which determines the relative importance of the attraction components, and G is a normalizing factor. The first component of T-VPack's second attraction function chooses critical-path LBs, and the second chooses LBs that share many connections with the LBs already packed into the cluster. By initializing and then packing clusters with

```
UnclusteredLBs = PatternMatchToLBs(LUTs,Registers);
LogicClusters = NULL;
while UnclusteredLBs != NULL do
    C = GetLBwithMostUsedInputs(UnclusteredLBs);
    while | C |< k do
        /*cluster is not full*/
        BestLB = MaxAttractionLegalLB(C,UnclusteredLBs);
        if BestLB == NULL then
            /*No LB can be added to this cluster*/
            break;
        endif
        UnclusteredLBs = UnclusteredLB − BestLB;
        C = C ∪ BestLB;
    endw
    if | C |< k then
        /*Cluster is not full - try hill climbing*/
        while | C |< k do
            BestLB = MinClusterInputIncreaseLB(C,UnclusteredLBs);
            C = C ∪ BestLB;
            UnclusteredLBs = UnclusteredLB − BestLB;
        endw
        if ClusterIsIllegal(C) then
            RestoreToLastLegalState(C,UnclusteredLBs);
        endif
    endif
    LogicClusters = LogicClusters ∪ C;
endw
```

Algorithm 2.1 Pseudo-code of the VPack Algorithm [22]

critical-path LBs, the algorithm is able to absorb long sequences of critical-path LBs into clusters. This minimizes circuit delay since the local interconnect within the cluster is significantly faster than the global interconnect of the FPGA. RPack [43] improves routability of a circuit by introducing a new set of routability metrics. RPack significantly reduced the channel widths required by circuits compared to VPack. T-RPack [43] is a timing driven version of RPack which is similar to T-VPack by giving added weight to grouping LBs on the critical path. iRAC [17] improves the routability of circuits even further by using an attraction function that attempts to encapsulate as many low fanout nets as possible within a cluster. If a net can be completely encapsulated within a cluster, there is no need to route that net in the external routing network. By encapsulating as many nets as possible within clusters, routability is improved because there are less external nets to route in total.

2.5.3.2 Top-down Approaches

The K-way partitioning problem seeks to minimize a given cost function of such an assignment. A standard cost function is net cut, which is the number of hyperedges that span more than one partition, or more generally, the sum of weights of such hyperedges. Constraints are typically imposed on the solution, and make the problem difficult. For example some vertices can be fixed in their parts or the total vertex weight in each part must be limited (balance constraint and FPGA clusters size). With balance constraints, the problem of partitioning optimally a hypergraph is known to be NP-hard [85]. However, since partitioning is critical in several practical applications, heuristic algorithms were developed with near-linear runtime. Such move-based heuristics for k-way hypergraph partitioning appear in [24, 34, 110].

Fiduccia-Mattheyses Algorithm

The Fiduccia-Mattheyses (FM) heuristics [34] work by prioritizing moves by gain. A move changes to which partition a particular vertex belongs, and the gain is the corresponding change of the cost function. After each vertex is moved, gains for connected modules are updated.

```
partitioning = initial_solution;
while solution quality improves do
    Initialize gain_container from partitioning;
    solution_cost = partitioning.get_cost();
    while not all vertices locked do
        move = choose_move();
        solution_cost += gain_container.get_gain(move);
        gain_container.lock_vertex(move.vertex());
        gain_update(move);
        partitioning.apply(move);
    endw
    roll back partitioning to best seen solution;
    gain_container.unlock_all();
endw
```

Algorithm 2.2 Pseudo-code for FM Heuristic [38]

The Fiduccia-Mattheyses (FM) heuristic for partitioning hypergraphs is an iterative improvement algorithm. FM starts with a possibly random solution and changes the solution by a sequence of moves which are organized as passes. At the beginning of a pass, all vertices are free to move (unlocked), and each possible move is labeled with the immediate change to the cost it would cause; this is called the gain of the move (positive gains reduce solution cost, while negative gains increase it). Iteratively, a move with highest gain is selected and executed, and the moving vertex is locked, i.e., is not allowed to move again during that pass. Since moving a vertex can change gains of adjacent vertices, after a move is executed all affected gains are updated. Selection and execution of a best-gain move, followed by gain update, are repeated until every vertex is locked. Then, the best solution seen during the pass is adopted as the starting solution of the next pass. The algorithm terminates when a

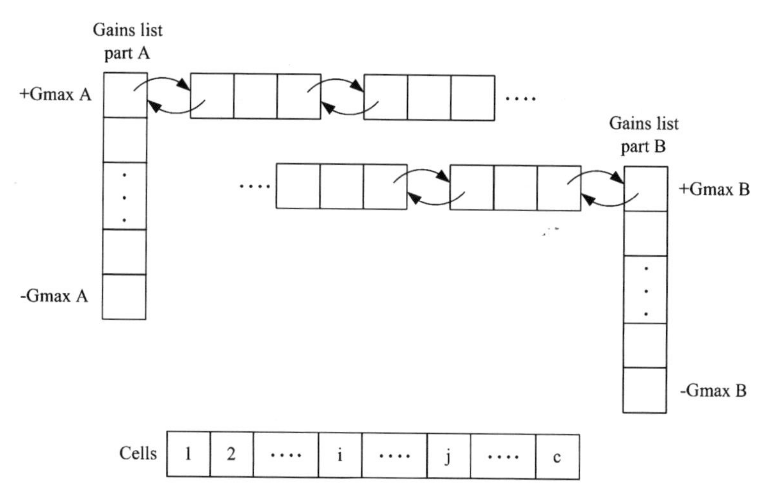

Fig. 2.21 The gain bucket structure as illustrated in [34]

pass fails to improve solution quality. Pseudo-code for the FM heuristic is given in algorithm 2.2.

The FM algorithm has 3 main components (1) computation of initial gain values at the beginning of a pass; (2) the retrieval of the best-gain (feasible) move; and (3) the update of all affected gain values after a move is made. One contribution of Fiduccia and Mattheyses lies in observing that circuit hypergraphs are sparse, and any move's gain is bounded between plus and minus the maximal vertex degree G_{max} in the hypergraph (times the maximal hyperedge weight, if weights are used). This allows prioritizing moves by their gains. All affected gains can be updated in amortized-constant time, giving overall linear complexity per pass [34]. All moves with the same gain are stored in a linked list representing a "gain bucket". Figure. 2.21 presents the gain bucket list structure. It is important to note that some gains G may be negative, and as such, FM performs hill-climbing and is not strictly greedy.

Multilevel Partitioning

The multilevel hypergraph partitioning framework was successfully verified by [31, 48, 49] and leads to the best known partitioning results ever since. The main advantage of multilevel partitioning over flat partitioners is its ability to search the solution space more effectively by spending comparatively more effort on smaller coarsened hypergraphs. Good coarsening algorithms allow for high correlation between good partitioning for coarsened hypergraphs and good partitioning for the initial hypergraph. Therefore, a thorough search at the top of the multilevel hierarchy is worthwhile because it is relatively inexpensive when compared to flat partitioning of the original hypergraph, but can still preserve most of the possible improvement.

The result is an algorithmic framework with both improved runtime and solution quality over a completely flat approach. Pseudo-code for an implementation of the multilevel partitioning framework is given in algorithm 2.3.

```
level = 0;
hierarchy[level] = hypergraph;
min_vertices = 200;
while hierarchy[level].vertex_count() > min_vertices do
    next_level = cluster(hierarchy[level]);
    level = level + 1;
    hierarchy[level] = next_level;
endw
partitioning[level] = a random initial solution for top-level hypergraph;
FM(hierarchy[level], partitioning[level]);
while level>0 do
    level = level - 1;
    partitioning[level] = project(partitioning[level+1], hierarchy[level]);
    FM(hierarchy[level], partitioning[level]);
endw
```

Algorithm 2.3 Pseudo-code for the Multilevel Partitioning Algorithm [38]

As illustrated in Fig. 2.22, multilevel partitioning consists of 3 main components: clustering, top-level partitioning and refinement or "uncoarsening". During clustering, hypergraph vertices are combined into clusters based on connectivity, leading to a smaller, clustered hypergraph. This step is repeated until obtaining only several hundred clusters and a hierarchy of clustered hypergraphs. We describe this hierarchy, as shown in Fig. 2.22, with the smaller hypergraphs being "higher" and the larger hypergraphs being "lower". The smallest (top-level) hypergraph is partitioned with a very fast initial solution generator and improved iteratively, for example, using the FM algorithm. The resulting partitioning is then interpreted as a solution for the next hypergraph in the hierarchy. During the refinement stage, solutions are projected from one level to the next and improved iteratively. Additionally, the hMETIS partitioning program [49] introduced several new heuristics that are incorporated into their multilevel partitioning implementation and are reportedly performance critical.

2.5.4 Placement

Placement algorithms determine which logic block within an FPGA should implement the corresponding logic block (instance) required by the circuit. The optimization goals consist in placing connected logic blocks close together to minimize the required wiring (wire length-driven placement), and sometimes to place blocks to balance the wiring density across the FPGA (routability-driven placement) or to maximize circuit speed (timing-driven placement). The 3 major classes of

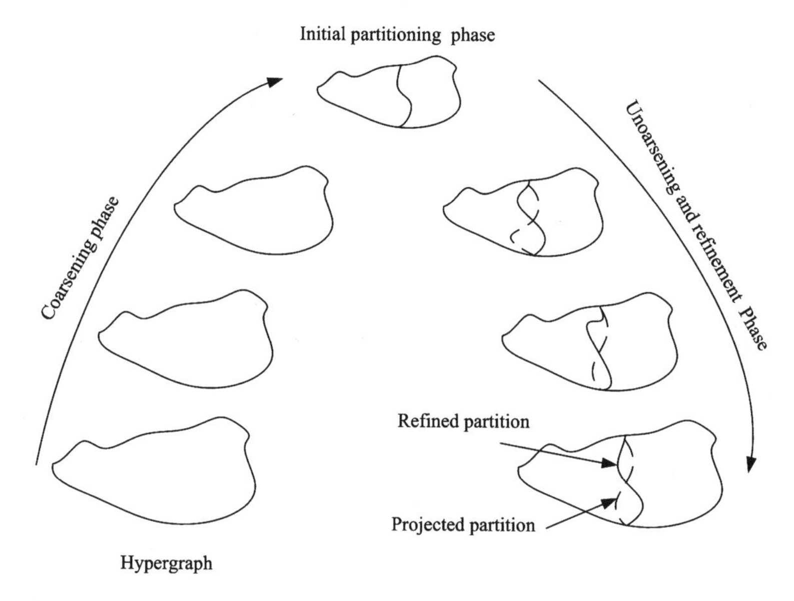

Fig. 2.22 Multilevel hypergraph bisection

placers in use today are min-cut (Partitioning-based) [6, 40], analytic [32, 53] which are often followed by local iterative improvement, and simulated annealing based placers [37, 105]. To investigate architectures fairly we must make sure that our CAD tools are attempting to use every FPGA's feature. This means that the optimization approach and goals of the placer may change from architecture to architecture. Partitioning and simulated annealing approaches are the most common and used in FPGA CAD tools. Thus we focus on both techniques in the sequel.

2.5.4.1 Simulated Annealing Based Approach

Simulated annealing mimics the annealing process used to cool gradually molten metal to produce high-quality metal objects [105]. Pseudo-code for a generic simulated annealing-based placer is shown in algorithm 2.4. A cost function is used to evaluate the quality of a given placement of logic blocks. For example, a common cost function in wirelength-driven placement is the sum over all nets of the half perimeter of their bounding boxes. An initial placement is created by assigning logic blocks randomly to the available locations in the FPGA. A large number of moves, or local improvements are then made to gradually improve the placement. A logic block is selected at random, and a new location for it is also selected randomly. The change in cost function that results from moving the selected logic block to the proposed new location is computed. If the cost decreases, the move is always accepted and the block is moved. If the cost increases, there is still a chance to accept the move, even though it makes the placement worse. This probability of acceptance is

```
S = RandomPlacement();
T = InitialTemperature();
R_limit = InitialR_limit;
while ExitCriterion() == false do
    while InnerLoopCriterion() == false do
        S_new = GenerateViaMove(S, R_limit);
        ΔC = Cost(S_new) − Cost(S);
        r = random(0,1);
        if r < e^(−ΔC/T) then
            S = S_new;
        endif
    endw
    T = UpdateTemp();
    R_limit = UpdateR_limit();
endw
```

Algorithm 2.4 Generic Simulated Annealing-based Placer [22]

given by $e^{-\frac{\Delta C}{T}}$, where ΔC is the change in cost function, and T is a parameter called temperature that controls probability of accepting moves that worsen the placement. Initially, T is high enough so almost all moves are accepted; it is gradually decreased as the placement improves, in such a way that eventually the probability of accepting a worsening move is very low. This ability to accept hill-climbing moves that make a placement worse allows simulated annealing to escape local minima of the cost function.

The R_{limit} parameter in algorithm 2.4 controls how close are together blocks must be to be considered for swapping. Initially, R_{limit} is fairly large, and swaps of blocks far apart on a chip are more likely. Throughout the annealing process, R_{limit} is adjusted to try to keep the fraction of accepted moves at any temperature close to 0.44. If the fraction of moves accepted, α, is less than 0.44, R_{limit} is reduced, while if α is greater than 0.44, R_{limit} is increased.

In [22], the objective cost function is a function of the total wirelength of the current placement. The wirelength is an estimate of the routing resources needed to completely route all nets in the netlist. Reductions in wirelength mean fewer routing wires and switches are required to route nets. This point is important because routing resources in an FPGA are limited. Fewer routing wires and switches typically are also translated into reductions of the delay incurred in routing nets between logic blocks. The total wirelength of a placement is estimated using a semi-perimeter metric, and is given by Eq. 2.2. N is the total number of nets in the netlist, $bbx(i)$ is the horizontal span of net i, $bby(i)$ is its vertical span, and $q(i)$ is a correction factor. Figure 2.23 illustrates the calculation of the horizontal and vertical spans of a hypothetical net that has 6 terminals.

$$WireCost = \sum_{i=1}^{N} q(i) \times (bb_x(i) + bb_y(i)) \quad (2.2)$$

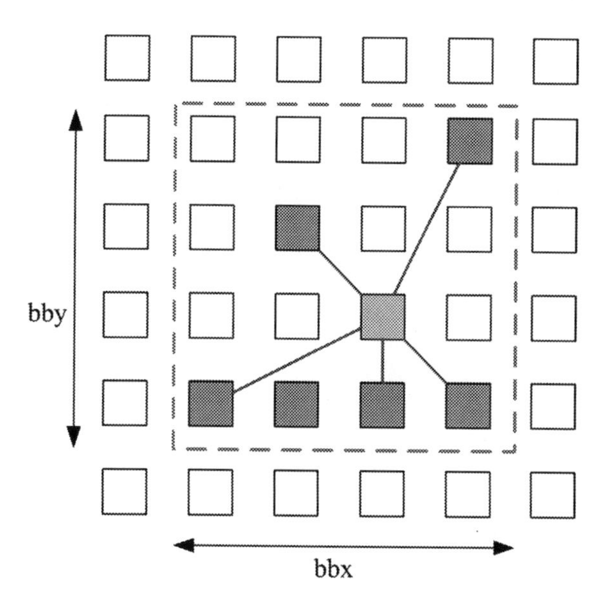

Fig. 2.23 Bounding box of a hypothetical 6-terminal net [22]

The temperature decrease rate, the exit criterion for terminating the anneal, the number of moves attempted at each temperature (InnerLoopCriterion), and the method by which potential moves are generated are defined by the annealing schedule. An efficient annealing schedule is crucial to obtain good results in a reasonable amount of CPU time. Many proposed annealing schedules are "fixed" schedules with no ability to adapt to different problems. Such schedules can work well within the narrow application range for which they are developed, but their lack of adaptability means they are not very general. In [86] authors propose an "adaptive" annealing schedule based on statistics computed during the anneal itself. Adaptive schedules are widely used to solve large scale optimization problems with many variables.

2.5.4.2 Partitioning Based Approach

Partitioning-based placement methods, are based on graph partitioning algorithms such as the Fiduccia-Mattheyses (FM) algorithm [34], and Kernighan Lin (KL) algorithm [6]. Partitioning-based placement are suitable to Tree-based FPGA architectures. The partitioner is applied recursively to each hierarchical level to distribute netlist cells between clusters. The aim is to reduce external communications and to collect highly connected cells into the same cluster.

The partitioning-based placement is also used in the case of Mesh-based FPGA. The device is divided into two parts, and a circuit partitioning algorithm is applied to determine the adequate part where a given logic block must be placed to minimize the number of cuts in the nets that connect the blocks between partitions, while leaving highly-connected blocks in one partition.

A divide-and-conquer strategy is used in these heuristics. By partitioning the problem into sub-parts, a drastic reduction in search space can be achieved. On the whole, these algorithms perform in the top-down manner, placing blocks in the general regions which they should belong to. In the Mesh FPGA case, partitioning-based placement algorithms are good from a "global" perspective, but they do not actually attempt to minimize wirelength. Therefore, the solutions obtained are sub-optimal in terms of wirelength. However, these classes of algorithms run very fast. They are normally used in conjunction with other search techniques for further quality improvement. Some algorithms [130] and [95] combine multi-level clustering and hierarchical simulated annealing to obtain ultra-fast placement with good quality. In the following chapters, the partitioning-based placement approach will be used only for Tree-based FPGA architectures.

2.5.5 Routing

The FPGA routing problem consists in assigning nets to routing resources such that no routing resource is shared by more than one net. $Pathfinder$ [80] is the current, state-of-the-art FPGA routing algorithm. $Pathfinder$ operates on a directed graph abstraction $G(V, E)$ of the routing resources in an FPGA. The set of vertices V in the graph represents the IO terminals of logic blocks and the routing wires in the interconnect structure. An edge between two vertices represents a potential connection between them. Figure 2.24 presents a part of a routing graph in a Mesh-based interconnect.

Given this graph abstraction, the routing problem for a given net is to find a directed tree embedded in G that connects the source terminal of the net to each of its sink terminals. Since the number of routing resources in an FPGA is limited, the goal of finding unique, non-intersecting trees for all the nets in a netlist is a difficult problem.

$Pathfinder$ uses an iterative, negotiation-based approach to successfully route all the nets in a netlist. During the first routing iteration, nets are freely routed without paying attention to resource sharing. Individual nets are routed using $Dijkstra$'s shortest path algorithm [111]. At the end of the first iteration, resources may be congested because multiple nets have used them. During subsequent iterations, the cost of using a resource is increased, based on the number of nets that share the resource, and the history of congestion on that resource. Thus, nets are made to negotiate for routing resources. If a resource is highly congested, nets which can use lower congestion alternatives are forced to do so. On the other hand, if the alternatives are more congested than the resource, then a net may still use that resource.

The cost of using a routing resource n during a routing iteration is given by Eq. 2.3.

$$c_n = (b_n + h_n) \times p_n \tag{2.3}$$

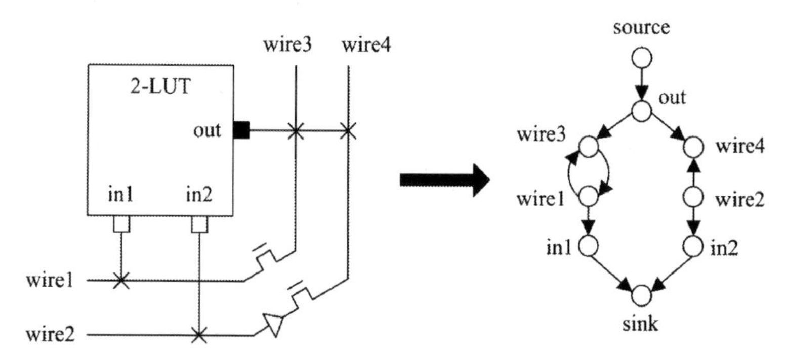

Fig. 2.24 Modeling FPGA architecture as a directed graph [22]

b_n is the base cost of using the resource n, h_n is related to the history of congestion during previous iterations, and p_n is proportional to the number of nets sharing the resource in the current iteration. The p_n term represents the cost of using a shared resource n, and the h_n term represents the cost of using a resource that has been shared during earlier routing iterations. The latter term is based on the intuition that a historically congested node should appear expensive, even if it is slightly shared currently. Cost functions and routing schedule were described in details in [22]. The Pseudo-code of the *Pathfinder* routing algorithm is presented in algorithm 2.5.

```
Let: RT_i be the set of nodes in the current routing of net i
while shared resources exist do
    /*Illegal routing*/
    foreach net, i do
        rip-up routing tree RT_i;
        RT(i) = s_i;
        foreach sink t_ij do
            Initialize priority queue PQ to RT_i at cost 0;
            while sink t_ij not found do
                Remove lowest cost node m from PQ;
                foreach fanout node n of node m do
                    Add n to PQ at PathCost(n) = c_n + PathCost(m);
                endfch
            endw
            foreach node n in path t_ij to s_i do
                /*backtrace*/
                Update c_n;
                Add n to RT_i;
            endfch
        endfch
    endfch
    update h_n for all n;
endw
```

Algorithm 2.5 Pseudo-code of the *Pathfinder* Routing Algorithm [80]

An important measure of routing quality produced by an FPGA routing algorithm is the critical path delay. The critical path delay of a routed netlist is the maximum delay of any combinational path in the netlist. The maximum frequency at which a netlist can be clocked has an inverse relationship with critical path delay. Thus, larger critical path delays slow down the operation of netlist. Delay information is incorporated into *Pathfinder* by redefining the cost of using a resource n (Eq. 2.4).

$$c_n = A_{ij} \times d_n + (1 - A_{ij}) \times (b_n + h_n) \times p_n \tag{2.4}$$

The c_n term is from Eq. 2.3, d_n is the delay incurred in using the resource, and A_{ij} is the criticality given by Eq. 2.5.

$$A_{ij} = \frac{D_{ij}}{D_{max}} \tag{2.5}$$

D_{ij} is the maximum delay of any combinational path going through the source and sink terminals of the net being routed, and D_{max} is the critical path delay of the netlist. Equation 2.4 is formulated as a sum of two cost terms. The first term in the equation represents the delay cost of using resource n, while the second term represents the congestion cost. When a net is routed, the value of A_{ij} determines whether the delay or the congestion cost of a resource dominates. If a net is near critical (i.e. its A_{ij} is close to 1), then congestion is largely ignored and the cost of using a resource is primarily determined by the delay term. If the criticality of a net is low, the congestion term in Eq. 2.4 dominates, and the route found for the net avoids congestion while potentially incurring delay.

Pathfinder has proved to be one of the most powerful FPGA routing algorithms to date. The negotiation-based framework that trades off delay for congestion is an extremely effective technique for routing signals on FPGAs. More importantly, *Pathfinder* is a truly architecture-adaptive routing algorithm. The algorithm operates on a directed graph abstraction of an FPGA's routing structure, and can thus be used to route netlists on any FPGA that can be represented as a directed routing graph.

2.5.6 Timing Analysis

Timing analysis [99] is used for two basic purposes:

- To determine the speed of circuits which have been completely placed and routed,
- To estimate the slack [68] of each source-sink connection during routing (placement and other parts of the CAD flow) in order to decide which connections must be made via fast paths to avoid slowing down the circuit.

First the circuit under consideration is presented as a directed graph. Nodes in the graph represent input and output pins of circuit elements such as LUTs, registers, and I/O pads. Connections between these nodes are modeled with edges in the graph. Edges are added between the inputs of combinational logic Blocks (LUTs) and their outputs. These edges are annotated with a delay corresponding to the physical delay between the nodes. Register input pins are not joined to register output pins. To determine the delay of the circuit, a breadth first traversal is performed on the graph starting at sources (input pads, and register outputs). Then the arrival time, $T_{arrival}$, at all nodes in the circuit is computed with the following equation:

$$T_{arrival}(i) = \max_{j \in fanin(i)}\{T_{arrival}(j) + delay(j, i)\}$$

where node i is the node currently being computed, and $delay(j, i)$ is the delay value of the edge joining node j to node i. The delay of the circuit is then the maximum arrival time, D_{max}, of all nodes in the circuit.
To guide a placement or routing algorithm, it is useful to know how much delay may be added to a connection before the path that the connection is on becomes critical. The amount of delay that may be added to a connection before it becomes critical is called the slack of that connection. To compute the slack of a connection, one must compute the required arrival time, $T_{required}$, at every node in the circuit. We first set the $T_{required}$ at all sinks (output pads and register inputs) to be D_{max}. Required arrival time is then propagated backwards starting from the sinks with the following equation:

$$T_{required}(i) = \min_{j \in fanout(i)}\{T_{required}(j) - delay(j, i)\}$$

Finally, the slack of a connection (i, j) driving node, j, is defined as:

$$Slack(i, j) = T_{required}(j) - T_{arrival}(i) - delay(i, j)$$

2.5.7 Bitstream Generation

Once a netlist is placed and routed on an FPGA, bitstream information is generated for the netlist. This bitstream is programmed on the FPGA using a bitstream loader. The bitstream of a netlist contains information as to which SRAM bit of an FPGA be programmed to 0 or to 1. The bitstream generator reads the technology mapping, packing and placement information to program the SRAM bits of Look-Up Tables. The routing information of a netlist is used to correctly program the SRAM bits of connection boxes and switch boxes.

2.6 Research Trends in Reconfigurable Architectures

Until now in this chapter a detailed overview of logic architecture, routing architecture and software flow of FPGAs is presented. In this section, we highlight some of the disadvantages associated with FPGAs and further we describe some of the trends that are currently being followed to remedy these disadvantages. FPGA-based products are basically very effective for low to medium volume production as they are easy to program and debug, and have less NRE cost and faster time-to-market. All these major advantages of an FPGA come through their reconfigurability which makes them general purpose and field programmable. But, the very same reconfigurability is the major cause of its disadvantages; thus making it larger, slower and more power consuming than ASICs.

However, the continued scaling of CMOS and increased integration has resulted in a number of alternative architectures for FPGAs. These architectures are mainly aimed to improve area, performance and power consumption of FPGA architectures. Some of these propositions are discussed in this section.

2.6.1 Heterogeneous FPGA Architectures

Use of hard-blocks in FPGAs improves their logic density. Hard-Blocks, in FPGAs increase their density, performance and power consumption. There can be different types of hard-blocks like multipliers, adders, memories, floating point units and DSP blocks etc. In this regard, [19] have incorporated embedded floating-point units in FPGAs, [30] have developed virtual embedded block methodology to model arbitrary embedded blocks on existing commercial FPGAs. Here some of the academic and commercial architectures are presented that make use of hard-blocks to improve overall efficiency of FPGAs.

2.6.1.1 Versatile Packing, Placement and Routing VPR

Versatile Packing, Placement and Routing for FPGAs (commonly known as VPR) [14, 22, 120] is the most widely used academic mesh-based FPGA exploration environment. It allows to explore mesh-based FPGA architectures by employing an empirical approach. Benchmark circuits are mapped, placed and routed on a desired FPGA architecture. Later, area and delay of FPGAs are measured to decide best architectural parameters. Different CAD tools in VPR are highly optimized to ensure high quality results.

Earlier version of VPR supported only homogeneous achitectures [120]. However, the latest version of VPR known as VPR 5.0 [81] supports hard-blocks (such as multiplier and memory blocks) and single-driver routing wires. Hard-blocks are restricted to be in one grid width column, and that column can be composed of only similar type of blocks. The height of a hard-block is quantized and it must be an integral multiple of grid units. In case a block height is indivisible with the height of FPGA, some grid locations are left empty. Figure 2.25 illustrates a heterogeneous FPGA with 8 different kinds of blocks.

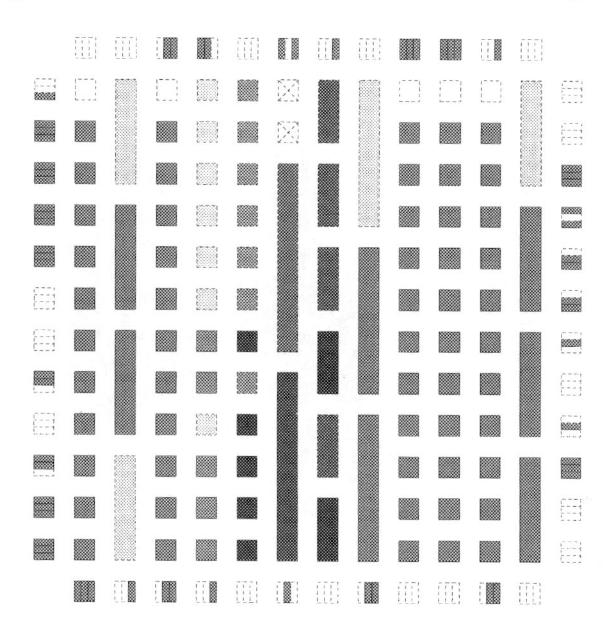

Fig. 2.25 A heterogeneous FPGA in VPR 5.0 [81]

2.6.1.2 Madeo, a Framework for Exploring Reconfigurable Architectures

Madeo [73] is another academic design suite for the exploration of reconfigurable architectures. It includes a modeling environment that supports multi-grained, heterogeneous architectures with irregular topologies. Madeo framework initially allows to model an FPGA architecture. The architecture characteristics are represented as a common abstract model. Once the architecture is defined, the CAD tools of Madeo are used to map a target netlist on the architecture. Madeo uses same placement and routing algorithms as used by VPR [120]. Along with placement and routing algorithms, it also embeds a bitstream generator, a netlist simulator, and a physical layout generator in its design suite. Madeo supports architectural prospection and very fast FPGA prototyping. Several FPGAs, including some commercial architectures (such as Xilinx Virtex family) and prospective ones (such as STMicro LPPGA) have been modeled using Madeo. The physical layout is produced as VHDL description.

2.6.1.3 Altera Architecture

Altera's Stratix IV [107] is an example of a commercial architecture that uses a heterogeneous mixture of blocks. Figure 2.26 shows the global architectural layout of Stratix IV. The logic structure of Stratix IV consists of LABs (Logic Array Blocks), memory blocks and digital signal processing (DSP) blocks. LABS are distributed symmetrically in rows and columns and are used to implement general purpose logic. The DSP blocks are used to implement full-precision multipliers of different

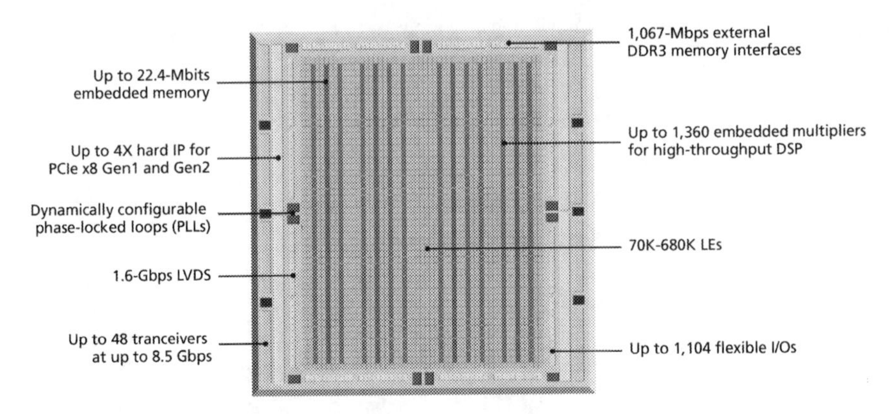

Fig. 2.26 Stratix IV architectural elements

granularities. The memory blocks and DSP blocks are placed in columns at equal distance with one another. Input and Output (I/Os) are located at the periphery of architecture.

Logic array blocks (LABs) and adaptive logic modules (ALMs) provide the basic logic capacity for Stratix IV device. They can be used to configure logic functions, arithmetic functions, and register functions. Each LAB consists of ten ALMs, carry chains, arithmetic chains, LAB control signals, local interconnect, and register chain connection lines. The local interconnect connects the ALMs that are inside same LAB. The direct link allows a LAB to drive into the local interconnect of its left or right neighboring LAB. The register chain connects the output of ALM register to the adjacent ALM register in the LAB. A memory LAB (MLAB) is a derivative of LAB which can be either used just like a simple LAB, or as a static random access memory (SRAM). Each ALM in an MLAB can be configured as a 64×1, or 32×2 blocks, resulting in a configuration of 64×10 or 32×20 simple dual-port SRAM block. MLAB and LAB blocks always coexist as pairs in Stratix IV families.

The DSP blocks in Stratix IV are optimized for signal processing applications such as Finite Impulse Response (FIR), Infinite Impulse Response (IIR), Fast Fourier Transform functions (FFT) and encoders etc. Stratix IV device has two to seven columns of DSP blocks that can implement different operations like multiplication, multiply-add, multiply-accumulate (MAC) and dynamic arithmetic or logical shift functions. The DSP block supports different multiplication operations such as 9×9, 12×12, 18×18 and 36×36 multiplication operations. The Stratix IV devices contain three different sizes of embedded SRAMs. The memory sizes include 640-bit memory logic array blocks (MLABs), 9-Kbit M9K blocks, and 144-Kbit M144K blocks. The MLABs have been optimized to implement filter delay lines, small FIFO buffers, and shift registers. M9K blocks can be used for general purpose memory applications, and M144K are generally meant to store code for a processor, packet buffering or video frame buffering.

2.6.2 FPGAs to Structured Architectures

The ease of designing and prototyping with FPGAs can be exploited to quickly design a hardware application on an FPGA. Later, improvements in area, speed, power and volume production can be achieved by migrating the application design from FPGA to other technologies such as Structured-ASICs. In this regard, Altera provides a facility to migrate its Stratix IV based application design to HardCopy IV [56]. Altera gives provision to migrate FPGA-based applications to Structured-ASIC. Their Structured-ASIC is called as HardCopy [56]. The main theme is to design, test and even initially ship a design using an FPGA. Later, the application circuit that is mapped on the FPGA can be seamlessly migrated to HardCopy for high volume production. Their latest HardCopy-IV devices offer pin-to-pin compatibility with the Stratix IV prototype, making them exact replacements for the FPGAs. Thus, the same system board and softwares developed for prototyping and field trials can be retained, enabling the lowest risk and fastest time-to-market for high-volume production. Moreover, when an application circuit is migrated from Stratix IV FPGA prototype to Hardcopy-VI, the core logic performance doubles and power consumption reduces by half.

The basic logic unit of HardCopy is termed as HCell. It is similar to Stratix IV logic cell (LAB) in the sense that the fabric consists of a regular pattern which is formed by tiling one or more basic cells in a two dimensional array. However, the difference is that HCell has no configuration memory. Different HCell candidates can be used, ranging from fine-grained NAND gates to multiplexors and coarse-grained LUTs. An array of such HCells, and a general purpose routing network which interconnects them is laid down on the lower layers of the chip. Specific layers are then reserved to form via connections or metal lines which are used to customize the generic array into specific functionality. Figure 2.27 illustrates the correspondence between an FPGA and a compatible structured ASIC. There is a one to one layout-level correspondence between MRAMs, phase-lock loops (PLLs), embedded memories, transceivers, and I/O blocks. The soft-logic DSP multipliers and logic cell fabric of the FPGA are re-synthesized to structured ASIC fabric. However, they remain functionally and electrically equivalent in FPGAs and HardCopy ASICs.

Apart from Altera, there are several other companies that provide a solution similar to that of Altera. For example, the eASIC Nextreme [41] uses an FPGA-like design flow to map an application design on SRAM programmable LUTs, which are later interconnected through mask programming of few upper routing layers. Tierlogic [113] is a recently launched FPGA vendor that offers 3D SRAM-based TierFPGA devices for prototyping and early production. The same design solution can be frozen to a TierASIC device with one low-NRE custom mask for error-free transition to an ASIC implementation. The SRAM layer is placed on an upper 3D layer of TierFPGA. Once the TierFPGA design is frozen, the bitstream information is used to create a single custom mask metal layer that will replace the SRAM programming layer.

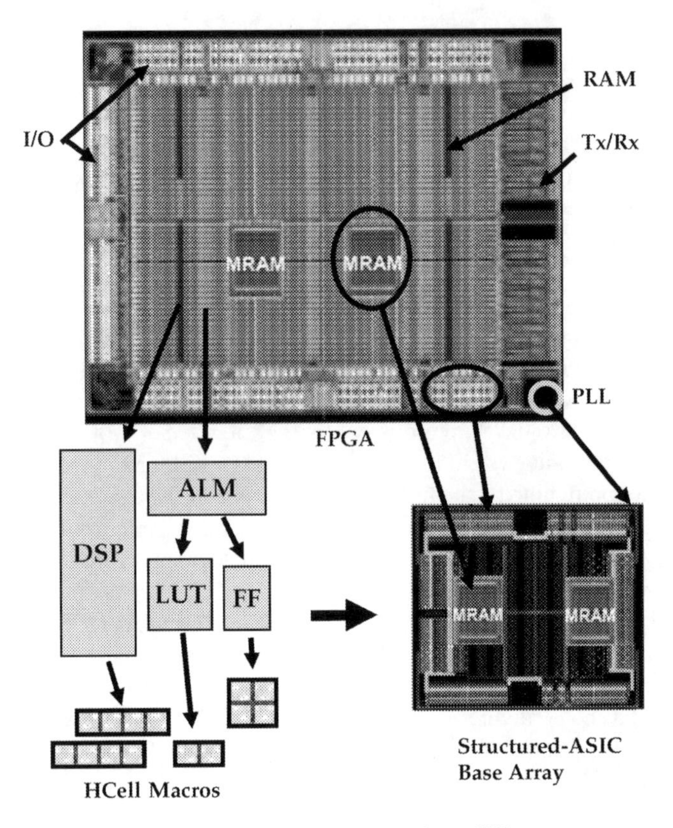

Fig. 2.27 FPGA/Structured-ASIC (HardCopy) Correspondence [59]

2.6.3 Configurable ASIC Cores

Configurable ASIC Core (cASIC) [35] is another example of reconfigurable devices that can implement a limited set of circuits which operate at mutually exclusive times. cASICs are intended as accelerator in domain-specific systems-on-a-chip, and are not designed to replace the entire ASIC-only chip. The host would execute software code, whereas compute-intensive sections can be executed on one or more cASICs. So, to execute the compute intensive sections, cASICs implement only data-path circuits and thus supports full-word blocks only (such as 16-bit wide multipliers, adders, RAMS, etc). Since the application domain of cASICs is more specific, they are significantly smaller than FPGAs. As hardware resources are shared between different netlists, cASICs are even smaller than the sum of the standard-cell based ASIC areas of individual circuits.

2.6.4 Processors Inside FPGAs

Considerable amount of FPGA area can be reduced by incorporating a microprocessor in an FPGA. A microprocessor can execute any less compute intensive task, whereas compute-intensive tasks can be executed on an FPGA. Similarly, a microprocessor based application can have huge speed-up gains if an FPGA is attached with it. An FPGA attached with a microprocessor can execute any compute intensive functionality as a customized hardware instruction. These advantages have compelled commercial FPGA vendors to provide microprocessor in their FPGAs so that complete system can be programmed on a single chip. Few vendors have integrated fixed hard processor on their FPGA (like AVR Processor integrated in Atmel FPSLIC [18] or PowerPC processors embedded in Xilinx Virtex-4 [126]). Others provide soft processor cores which are highly optimized to be mapped on the programmable resources of FPGA. Altera's Nios [90] and Xilinx's Microblaze [88] are soft processor meant for FPGA designs which allow custom hardware instructions. [96] have shown that considerable area gains can be achieved if these soft processors for FPGAs are optimized for particular applications. They have shown that unused instructions in a soft processor can be removed and different architectural tradeoffs can be selected to achieve on average 25% area gain for soft processors required for specific applications. Reconfigurable units can also be attached with microprocessors to achieve execution time speedup in software programs. [28, 70, 104] have incorporated a reconfigurable unit with microprocessors to achieve execution-time speedup.

2.6.5 Application Specific FPGAs

The type of logic blocks and the routing network in an FPGA can be optimized to gain area and performance advantages for a given application domain (controlpath-oriented applications, datapath-oriented applications, etc). These types of FPGAs may include different variety of desired hard-blocks, appropriate amount of flexibility required for the given application domain or bus-based interconnects rather than bit-based interconnects. Authors in [83] have presented a reconfigurable arithmetic array for multimedia applications which they call as CHESS. The principal goal of CHESS was to increase arithmetic computational density, to enhance the flexibility, and to increase the bandwidth and capacity of internal memories significantly beyond the capabilities of existing commercial FPGAs. These goals were achieved by proposing an array of ALUs with embedded RAMs where each ALU is 4-bit wide and supports 16 instructions. Similarly, authors in [42] present a coarse-grained, field programmable architecture for constructing deep computational pipelines. This architecture can efficiently implement applications related to media, signal processing, scientific computing and communications. Further, authors in [128] have used bus-based routing and logic blocks to improve density of FPGAs for datapath circuits. This is a partial multi-bit FPGA architecture that is designed to exploit the regularity that most of the datapath circuits exhibit.

2.6.6 Time-Multiplexed FPGAs

Time-multiplexed FPGAs increase the capacity of FPGAs by executing different portions of a circuit in a time-multiplexed mode [89, 114]. An application design is divided into different sub-circuits, and each sub-circuit runs as an individual context of FPGA. The state information of each sub-circuit is saved in context registers before a new context runs on FPGA. Authors in [114] have proposed a time-multiplexed FPGA architecture where a large circuit is divided into sub-circuits and each sub-circuit is sequentially executed on a time-multiplexed FPGA. Such an FPGA stores a set of configuration bits for all contexts. A context is shifted simply by using the SRAM bits dedicated to a particular context. The combinatorial and sequential outputs of a sub-circuit that are required by other sub-circuits are saved in context registers which can be easily accessed by sub-circuits at different times.

Time-Multiplexed FPGAs increase their capacity by actually adding more SRAM bits rather than more CLBs. These FPGAs increase the logic capacity by dynamically reusing the hardware. The configuration bits of only the currently executing context are active, the configuration bits for the remaining supported contexts are inactive. Intermediate results are saved and then shared with the contexts still to be run. Each context takes a micro-cycle time to execute one context. The sum of the micro-cycles of all the contexts makes one user-cycle. The entire time-multiplexed FPGA or its smaller portion can be configured to (i) execute a single design, where each context runs a sub-design, (ii) execute multiple designs in time-multiplexed modes, or (iii) execute statically only one single design. Tabula [109] is a recently launched FPGA vendor that provides time-multiplexed FPGAs. It dynamically reconfigures logic, memory, and interconnect at multi-GHz rates with a Spacetime compiler.

2.6.7 Asynchronous FPGA Architecture

Another alternative approach that has been proposed to improve the overall performance of FPGA architecture is the use of asynchronous design elements. Conventionally, digital circuits are designed for synchronous operation and in turn FPGA architectures have focused primarily on implementing synchronous circuits. Asynchronous designs are proposed to improve the energy efficiency of asynchronous FPGAs since asynchronous designs offer potentially lower energy as energy is consumed only when necessary. Also the asynchronous architectures can simplify the design process as complex clock distribution networks become unnecessary.

The first asynchronous FPGA was developed by [57]. It consisted the modified version of previously developed synchronous FPGA architecture. Its logic block was similar to the conventional logic block with added features of fast feedback and a latch that could be used to initialize an asynchronous circuit. Another asynchronous architecture was proposed in [112]. This architecture is designed specifically for dataflow applications. Its logic block is similar to that of synchronous architecture, along with it consists of units such as split unit which enables conditional forwarding of data and a merge unit that allows for conditional selection of data from different sources. An alternative to fully asynchronous design is a globally asynchronous, locally synchronous approach (GALS). This approach is used by [69] where authors have introduced a level of hierarchy into the FPGA architecture. Standard hard or soft synchronous logic blocks are grouped together to form large synchronous blocks and communication between these blocks is done asynchronously. More recently, authors in [131] have applied the GALS approach on Network on Chip architectures to improve the performance, energy consumption and the yield of future architectures in a synergistic manner.

It is clear that, despite each architecture offering its own benefits, a number of architectural questions remain unresolved for asynchronous FPGAs. Many architectures rely on logic blocks similar to those used for synchronous designs [57, 69] and, therefore, the same architectural issues such as LUT size, cluster size, and routing topology must be investigated. In addition to those questions, asynchronous FPGAs also add the challenge of determining the appropriate synchronization methodology.

2.7 Summary and Conclusion

In this chapter initially a brief introduction of traditional logic and routing architectures of FPGAs is presented. Later, different steps involved in the FPGA design flow are detailed. Finally various approaches that have been employed to reduce few disadvantages of FPGAs and ASICs, with or without compromising their major benefits are described. Figure 2.28 presents a rough comparison of different solutions used to reduce the drawbacks of FPGAs and ASICs. The remaining chapters of this book will focus on the exploration of tree-based FPGA architectures using hard-blocks, tree-based application specific Inflexible FPGAs (ASIF), and their automatic layout generation methods.

This book presents new environment for the exploration of tree-based heterogeneous FPGAs. This environment is used to explore different architecture techniques for tree-based heterogeneous FPGA architecture. This book also presents an optimized environment for mesh-based heterogeneous FPGA. Further, the environments of two architectures are evaluated through the experimental results that are obtained by mapping a number of heterogeneous benchmarks on the two architectures.

Altera [11] has proposed a new idea to prototype, test, and even ship initial few designs on an FPGA, later the FPGA based design can be migrated to Structured-ASIC (known as HardCopy). However, migration of an FPGA-based product to Structured-ASIC supports only a single application design. An ASIF retains this

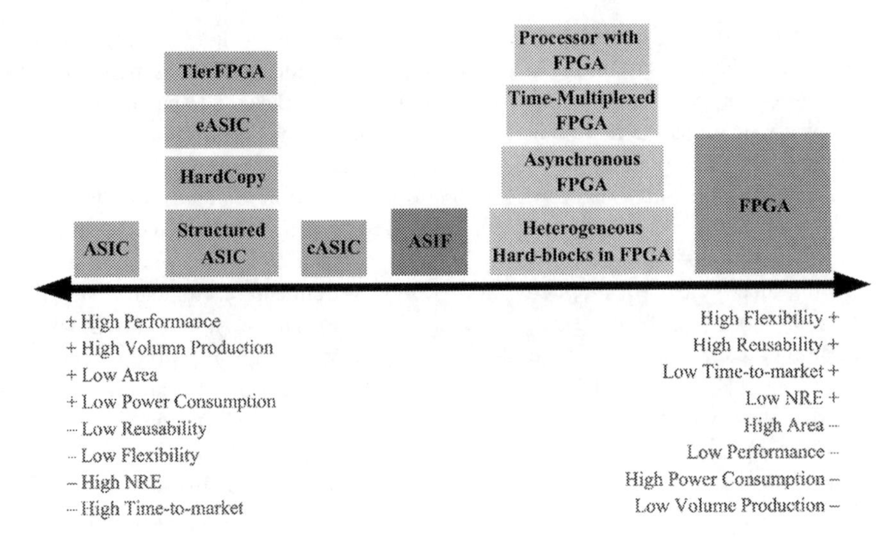

Fig. 2.28 Comparison of different solutions used to reduce ASIC and FPGA drawbacks

property, and can be a possible future extension for the migration of FPGA-based applications to Structured-ASIC. Thus when an FPGA-based product is in the final phase of its development cycle, and if the set of circuits to be mapped on the FPGA are known, the FPGA can be reduced to an ASIF for the given set of application designs. This book presents a new tree-based ASIF and a detailed comparison of tree-based ASIF is performed with mesh-based ASIF. This book also presents automatic layout generation techniques for domain-specific FPGA and ASIF architectures.

Chapter 3
Homogeneous Architectures Exploration Environments

The advancement in process technology has greatly enhanced the capacity of FPGAs and they have become increasingly popular for the implementation of larger designs. Design of large devices implies fundamental and efficient innovation in FPGA architecture to improve density, speed and power optimization of the architecture. It has been observed that most of the designs exhibit a locality in their connections and specific architectures are required to be designed to exploit their locality and improve the architecture efficiency. In this chapter we present an exploration environment for a new tree-based homogeneous FPGA architecture [132] that exploits the locality that most of the designs exhibit. Routability and interconnect area of this architecture depends on switch boxes topology and signals bandwidth (in/out signals per cluster). In tree-based FPGA we use full crossbar switch boxes and we aim at exploiting the available flexibility to reduce signals bandwidth based on suitable partitioning approaches.

It is well established that the quality of an FPGA based implementation is largely dependant on the accompanying flow that is used to map different applications on the FPGA architecture. Benefits of an otherwise well designed, feature rich FPGA architecture might be impaired if the CAD tools can not take advantage of the features that FPGA provides. In this book we have designed an environment for tree-based architecture that exploits the features of this architecture. The environment is based on a mixture of generalized and specifically developed tools. These tools are used to map different application on the tree-based architecture. A reference mesh-based architecture and its associated exploration environment is also presented in this chapter and the two architectures are compared using the results that are obtained through the exploration environments of both architectures.

U. Farooq et al., *Tree-Based Heterogeneous FPGA Architectures*,
DOI: 10.1007/978-1-4614-3594-5_3,

3.1 Reference FPGA Architectures

This section gives basic overview of the two FPGA architectures that are used in this book. A brief overview of generalized mesh and tree-based architectures is already presented in Chap. 2. Here, we present further details and also present the customized software flow that we have developed for both architectures. Although two architectures are comprised of similar logic and routing resources (i.e. configurable logic blocks, multiplexors, configuration memory etc), it is the arrangement of these resources that differentiates the two architectures. In a tree-based architecture logic and routing resources are arranged in hierarchical manner while in mesh-based architecture resources are arranged in an island-style.

3.1.1 Mesh-Based FPGA Architecture

A mesh-based FPGA is represented as a grid of equally sized slots which is termed as slot-grid. The reference mesh-based FPGA is a VPR-style (Versatile Place & Route) [22] architecture, as shown in Fig. 3.1. It contains Configurable Logic Blocks (CLBs) arranged on a two dimensional grid. Each CLB contains one Look-Up Table with c_{in} inputs and $c_{out} = 1$ output and one Flip-Flop (FF). In mesh-based FPGA, input and output pads are arranged at the periphery of the slot-grid as shown in Fig. 3.1 and a CLB is surrounded by a single-driver unidirectional routing network [77]. The routing network is arranged in the form of uniform horizontal and vertical routing channels where each channel contains a fixed number of routing tracks which is termed as the channel width of the architecture.

A mesh-based FPGA is divided into "tiles" that are repeated horizontally and vertically to form a complete FPGA. A CLB, along with its surrounding routing network, forms the tile of the architecture which is repeated horizontally and vertically to form the complete FPGA architecture. A single FPGA tile, surrounded by its routing network is shown in Fig. 3.2. In this figure CLB contains a LUT which has 4 inputs and 1 output. Each of the 4 inputs of a CLB are connected to 4 adjacent routing channels. The output pin of a CLB connects with the routing channel on its top and right through the diagonal connections of the switch box (highlighted in the bottom-left switch box shown in Fig. 3.2). The switch box uses unidirectional, disjoint topology to connect different routing tracks together. The connectivity of a routing track incident on a switch block with routing tracks of other routing channels that are incident on the same switch block, termed as switch block flexibility (Fs), is set to be 3. The connectivity of routing channel with the input and output pins of a CLB, abbreviated as Fc(in) and Fc(out), is set to be maximum at 1.0. The channel width is varied according to the netlist requirement but remains in multiples of 2 [77].

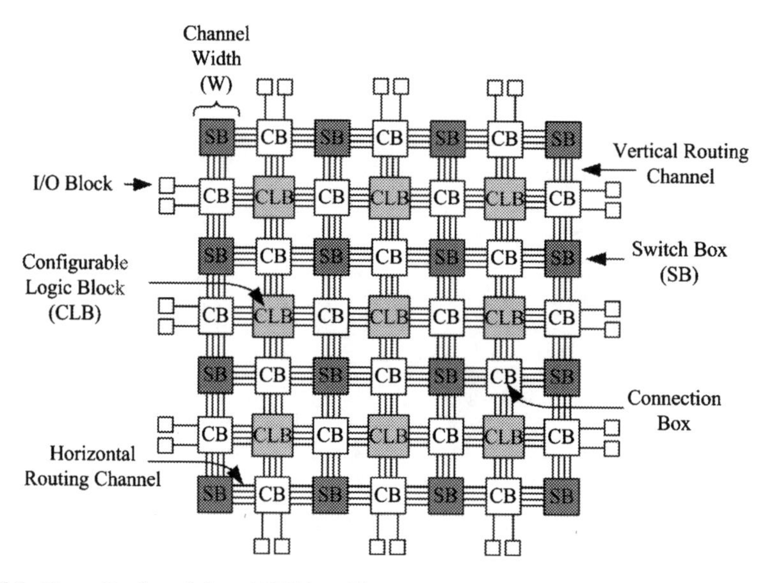

Fig. 3.1 Generalized mesh-based FPGA architecture

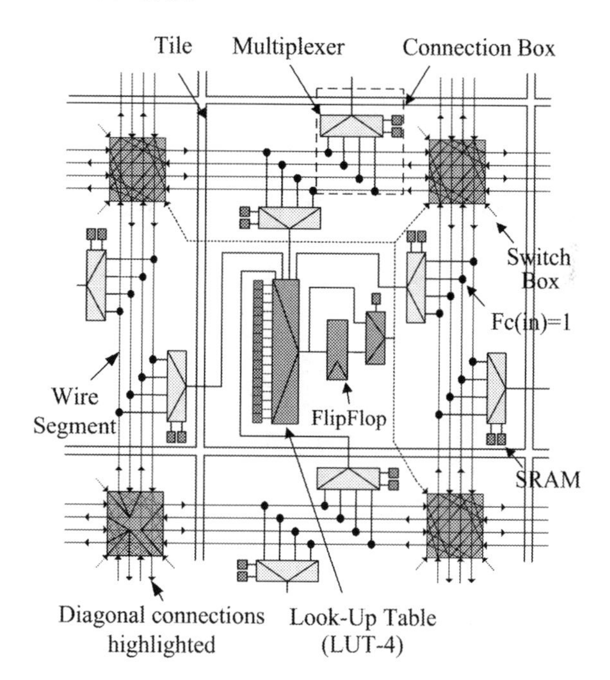

Fig. 3.2 Detailed interconnect of a CLB with its surrounding routing network

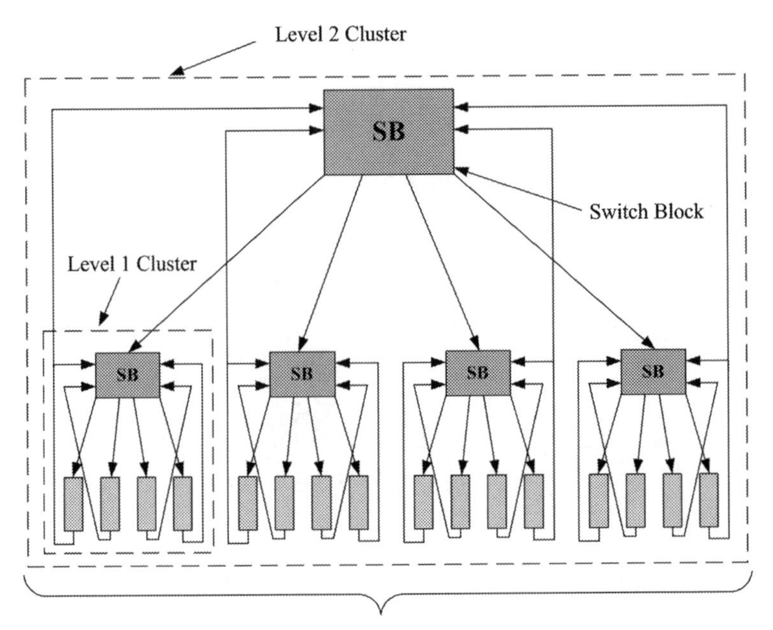

Fig. 3.3 Generalized tree-based FPGA architecture

3.1.2 Tree-Based FPGA Architecture

In a tree-based architecture logic and routing resources are partitioned into a multilevel clustered structure where each cluster contains sub-clusters and signals coming in and out of clusters communicate with other clusters using switch blocks. Generalized example of a two level, arity 4 tree-based architecture is shown in Fig. 3.3 where level 2 cluster contains 4 level 1 sub-clusters and each level 1 sub-cluster contains 4 CLBs. Since each cluster contains 4 sub-clusters in this architecture, it is termed here as arity 4 architecture. Tree-based architecture uses single driver unidirectional wires as bidirectional wires introduce considerable routing and area overhead [77]. Just like [3, 4], for tree-based architecture, a fully hierarchical interconnect is built where inter level signal bandwidth grows according to Rent's rule [25]. As it can be seen from Fig. 3.3 that contrary to [4], our tree-based architecture is divided into two unidirectional routing networks: the downward network and upward network. The downward network is inspired from SPIN [97] and it uses butterfly fat tree topology [33] to connect different signals using switch boxes and unidirectional wires. The upward network uses hierarchy and connects different signals using linearly populated switch boxes and unidirectional wires.

3.1.2.1 Architecture Interconnect Details

In a tree-based architecture, each configurable logic block (CLB) contains one Look-Up-Table (LUT) with c_{in} inputs and $c_{out} = 1$ output, followed by a bypass Flip-Flop. CLBs are grouped into k sized clusters and interconnect is organized into levels. For example in Fig. 3.3, the cluster size is 4, architecture has two levels and it supports 16 CLBs. Let $nb\ell$ denote the number of levels of a given Tree containing N leaves ($nb\ell = log_k(N)$). In each level ℓ we have $\frac{N}{k^\ell}$ clusters; C is the set of clusters in all levels. A cluster with index c belonging to level ℓ is noted by $cluster(\ell, c)$. A cluster switch block is divided into separate Downward Mini Switch Boxes (DMSBs) and Upward Mini Switch Boxes (UMSBs). DMSBs are responsible for downward interconnect and UMSBs are responsible for upward interconnect. DMSBs and UMSBs are combined together to route different signals of the netlists that are mapped on the architecture. These DMSBs and UMSBs are unidirectional full cross bar switches that connect signals coming into the cluster to its sub-clusters and signals going out of a cluster to the other clusters of hierarchy. Each $cluster(\ell, c)$ where $\ell \geq 1$ contains a set of inputs $N_{in}(\ell)$, a set of outputs $N_{out}(\ell)$, a set of downward and upward switch boxes and k sub-clusters. The inputs and outputs of cluster $cluster(\ell, c)$ are divided equally among its DMSBs and UMSBs which are used to connect these inputs and outputs to its sub-clusters and to other clusters of hierarchy. Sub-clusters of $cluster(\ell, c)$ are $cluster(\ell - 1, k.c + i)$ where $i \in \{0, 1, 2, \ldots, k - 1\}$. k is called $cluster(\ell, c)$ arity.

Each cluster in level 0 is denoted $cluster(0, c)$ or $leafcluster(c)$ and corresponds to the Configurable Logic Block (CLB) and contains c_{in} inputs, 1 output, no MSBs and no sub-cluster. Each $cluster(\ell, c)$ where $\ell < nb\ell - 1$ has an owner in level ℓ', where $\ell' > \ell$, denoted $cluster(\ell', c \div k^{(\ell'-\ell)})$. We define for each $cluster(\ell, c)$ a position inside its owner in level $\ell + 1$ (direct owner) by the following function:

$$\begin{aligned} pos: \quad & C \longrightarrow \{0, 1, 2, \ldots, k - 1\} \\ & cluster(\ell, c) \longmapsto c \ mod \ k \end{aligned}$$

Two clusters belonging to level ℓ and with the same owner at level $\ell + 1$ have two different positions. To get the cluster owner in level ℓ' of $cluster(\ell, c)$ ($\ell < \ell' \leq nb\ell - 1$) we define the function:

$$\begin{aligned} owner: \quad & C \times \mathbb{N} \longrightarrow C \\ & (cluster(\ell, c), \ell') \longmapsto cluster(\ell', c \div k^{\ell'-\ell}) \end{aligned}$$

3.1.2.2 Downward Network

Figure 3.4 shows a sparse downward network based on unidirectional DMSBs. The downward interconnect topology is similar to the butterfly fat tree. Each DMSB of a $cluster(\ell, c)$ where $\ell > 1$ is connected to each sub-cluster through one and only

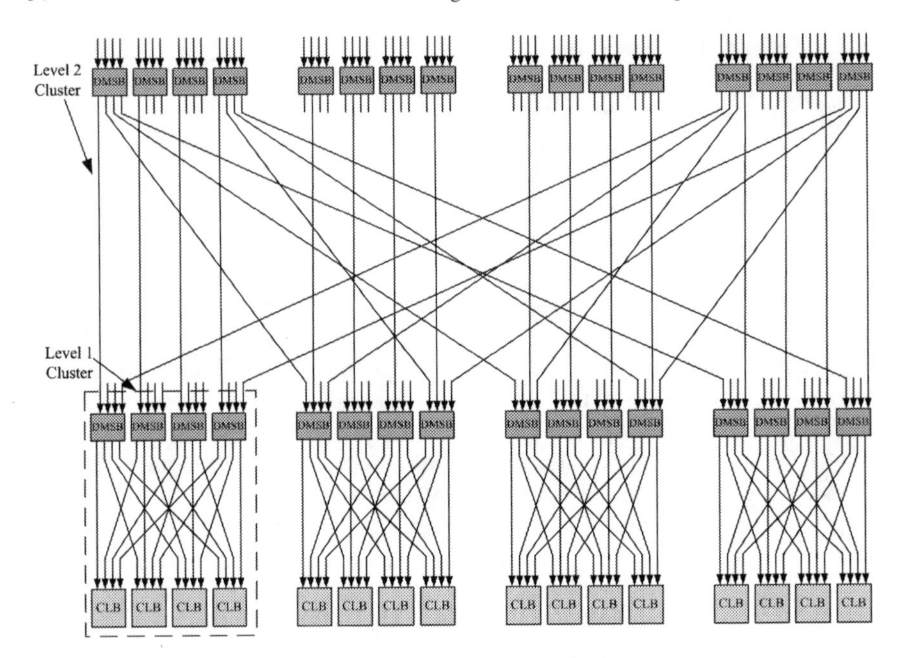

Fig. 3.4 Detailed downward interconnect of a tree-based architecture

one input pin. Thus, the DMSBs number in a cluster situated in level ℓ is equal to the input number of a cluster situated in level $\ell - 1$: $nbDMSB(\ell) = N_{in}(l - 1)$. For example in Fig. 3.4, the number of DMSBs at level 2 are equal to the number of inputs of a sub-cluster at level 1 (i.e. 16).

We name *DMSB*(ℓ, c, m) as the successor of $DMSB(\ell', c', m')$ where $0 < \ell < \ell'$ if there is a downward directed path from $DMSB(\ell', c', m')$ to $DMSB(\ell, c, m)$. The path between a DMSB and its successor is unique. Thus each $DMSB(\ell', c', m')$ has a successor in each sub-cluster belonging to level ℓ $DMSB(\ell, c, m)$ where $0 < \ell < \ell'$.

3.1.2.3 Upward Network

Figure 3.5 shows the upward network of a tree-based architecture that uses UMSBs that allow LBs outputs to reach a larger number of Downward MSBs (DMSBs). The UMSBs are organized in a way that allows CLBs belonging to the same "owner cluster" (at level 1 or above) to reach exactly the same set of DMSBs at each level. The interconnect offers more routing paths to connect a net source to a given sink. In this case we are more likely to achieve highly congested netlists routing. This gives an efficient solution for mapping netlists since instances may have different fanout sizes. For example in Fig. 3.5, a CLB output can reach all 4 DMSBs of its owner cluster at level 1 and all the 16 DMSBs of its owner cluster at level 2.

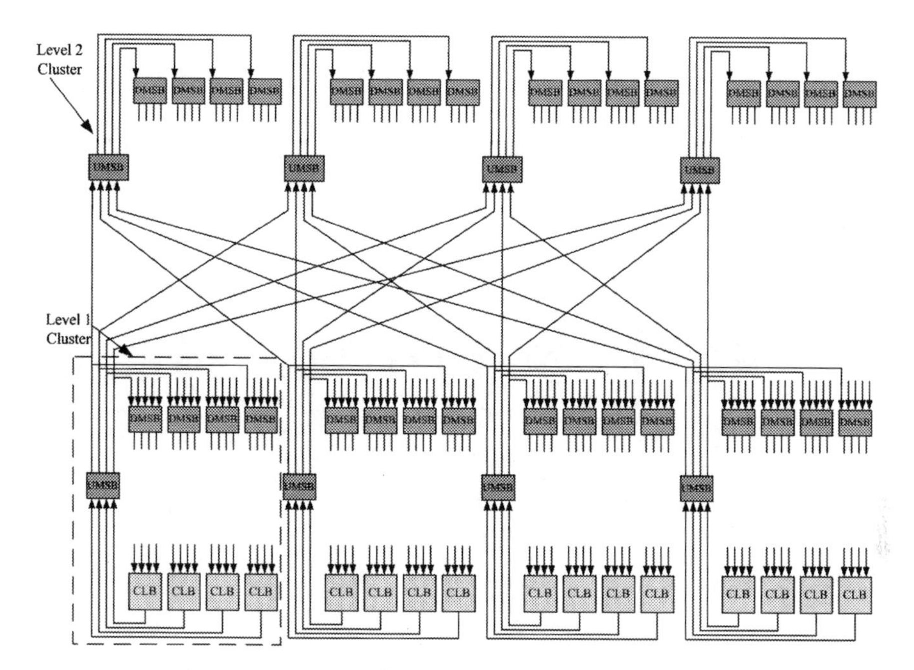

Fig. 3.5 Detailed upward interconnect of a tree-based architecture

3.1.2.4 I/O Connections

Figure 3.6 shows the combined downward and upward interconnect of tree-based architecture. Also it can be seen from this figure that output and input pads are grouped into specific clusters. The cluster size and the level where it is located can be modified to obtain the best design fit. Each input pad is connected to all UMSBs of the upper level. In this way each input pad can reach all CLBs of the architecture with different paths.

Similarly, output pads are connected to all DMSBs of the upper level; in this way they can be reached from all CLBs through different paths. The flexibility of I/O pads is kept higher than those of CLBs to ensure the routing of highly congested netlists.

3.1.2.5 Interconnect Depopulation

Although the use of DMSBs and UMSBs gives the architecture a great amount of flexibility in terms of the number of paths that can be used to route different signals on the architecture, it increases the number of switches in the architecture which can increase the area of the architecture. This can be compensated by reduction of in/out signals bandwidth of clusters at every level. In fact Rent's rule [25] is easily adapted to tree-based structure:

$$IO = c.k^{\ell.p} \tag{3.1}$$

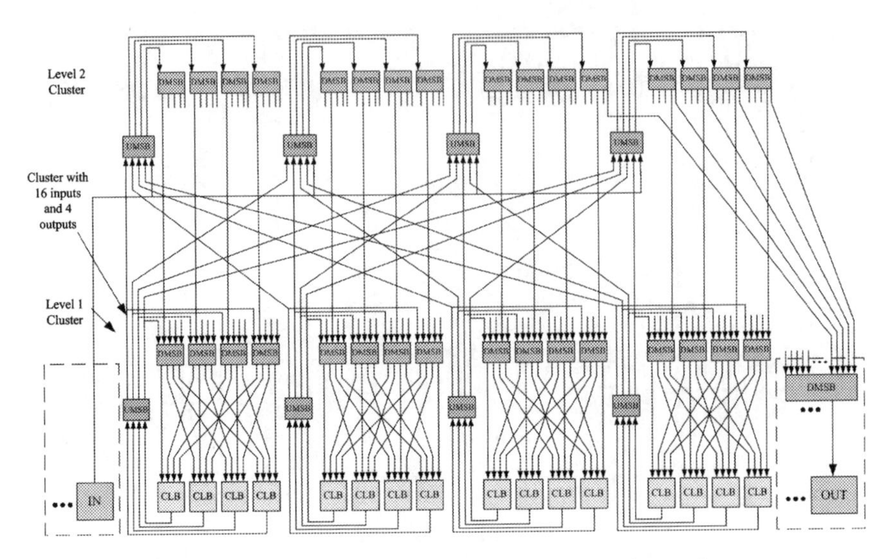

Fig. 3.6 Tree-based architecture with detailed *downward* and *upward* networks

where ℓ is a Tree level, k is the cluster arity, c is the number of in/out pins of a logic block and IO the number of in/out pins of a cluster situated at level ℓ. Intuitively, p represents the locality in interconnect requirements. If most connections are purely local and only few of them come in from the exterior of a local region, p will be small. In Tree-based architecture, both the upward and downward interconnects populations depend on this parameter. We can depopulate the routing interconnect by reducing the value of p which in turn reduces the signal bandwidth of the architecture. By doing so the architecture routability is reduced too. An example of a depopulated tree-based interconnect is shown in Fig. 3.7. Compared to the example shown in Fig. 3.6, number of inputs of each cluster at level 1 are reduced from 16 to 10 and the number of outputs are reduced from 4 to 3. By reducing the inputs and outputs the number of switches are reduced by 21% and value of p is reduced from 1 to 0.79. Although, this reduction improves the area of the architecture, it reduces its flexibility too. Thus we have to find the best tradeoff between interconnect population and logic blocks occupancy. Dehon showed in [4] that the best way to improve circuit density is to balance logic blocks and interconnect utilization. In tree-based architecture, the logic occupancy factor is controlled by N, the leaves (CLBs) number in the Tree. N is directly related to the number of levels and the clusters arity k. In most cases N is larger than the number of netlists instances. This means that in these cases we have a low logic utilization. This is not really penalizing since it can be compensated by a depopulated interconnect. In other words, the area overhead due to unused CLBs is compensated by congestion spreading and interconnect reduction.

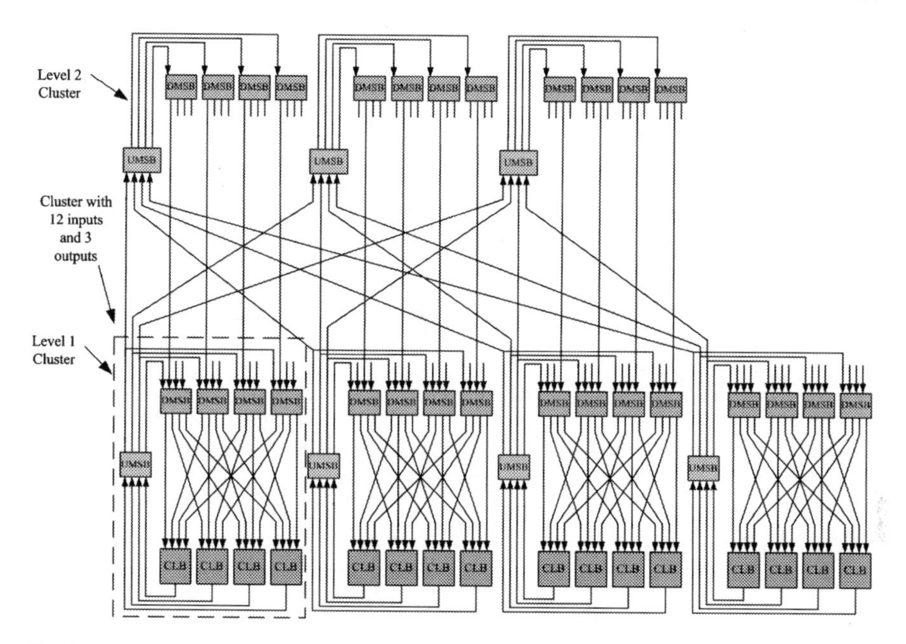

Fig. 3.7 Depopulation of tree-based architecture using Rent's rule

3.1.2.6 Rent's Rule Based Model

Based on Rent's rule presented in Eq. (3.1), we evaluate the Tree architecture switches requirement to connect LBs.

Switches requirement

We model upward and downward networks separately:

Downward network:

We note:

- $N_{in}(\ell)$ the number of inputs of a cluster located at level ℓ.
- $N_{out}(\ell)$ the number of outputs of a cluster located at level ℓ.
- c_{out} the number of outputs of an LB.
- c_{in} the number of inputs of an LB.
- k clusters arity (size).

Clusters located at level ℓ contain $N_{in}(\ell - 1)$ DMSB with k outputs and $\frac{N_{in}(\ell)+kN_{out}(\ell-1)}{N_{in}(\ell-1)}$ inputs. As we assume that the DMSB are full crossbar devices, we get $k(N_{in}(\ell) + kN_{out}(\ell - 1))$ switches in the switch box of a level ℓ cluster. Since we have $\frac{N}{k^\ell}$ clusters in level ℓ, we get a total number of switches, related to the downward network, given by:

$$\sum_{\ell=1}^{\log_k(N)} k \times N \times \frac{N_{in}(\ell) + kN_{out}(\ell-1)}{k^\ell}$$

$N_{out}(0) = c_{out}$ is the number of outputs of a Basic Logic Block. Following Eq. (3.1), we get $N_{in}(\ell) = c_{in}.k^{\ell.p}$ and $N_{out}(\ell-1) = c_{out}.k^{(\ell-1)p}$. The total number of switches used in the downward network is:

$$N_{switch}(down) = N \times (k^p c_{in} + kc_{out}) \times \sum_{\ell=1}^{\log_k(N)} k^{(p-1)(\ell-1)}$$

Upward network:

Clusters located at level ℓ contain $N_{out}(\ell-1)$ UMSB with k inputs and k outputs. As we assume that UMSB are full crossbar devices, we get $k^2 \times N_{out}(\ell-1)$ switches in the switch box of a level ℓ cluster. As we have $\frac{N}{k^\ell}$ clusters at level ℓ we get the total number of switches, related to the upward network:

$$\sum_{\ell=1}^{\log_k(N)} \frac{k^2 \times N}{k^\ell} \times N_{out}(\ell-1)$$

$N_{out}(0) = c_{out}$ is the number of outputs of a Basic Logic Block. Following (3.1), we get $N_{out}(\ell-1) = c_{out}.k^{(\ell-1)p}$.
The total number of switches used in the upward interconnect is:

$$N_{switch}(up) = N \times k \times c_{out} \times \sum_{\ell=1}^{\log_k(N)} k^{(p-1)(\ell-1)}$$

The total number of Tree-based interconnect switches is

$$N_{switch}(Tree) = N_{switch}(down) + N_{switch}(up)$$

$$N_{switch}(Tree) = N \times (k^p c_{in} + 2kc_{out}) \times \sum_{\ell=1}^{\log_k(N)} k^{(p-1)(\ell-1)}$$

The number of switches per Logic Block is:

$$N_{switch}(LB) = (k^p c_{in} + 2kc_{out}) \times \sum_{\ell=1}^{\log_k(N)} k^{(p-1)(\ell-1)}$$

$$N_{switch}(LB) = \begin{cases} (k^p c_{in} + 2kc_{out}) \times \frac{1-N^{p-1}}{1-k^{p-1}} \text{ if } p \neq 1 \\ (k^p c_{in} + 2kc_{out}) \times \log_k(N) \text{ if } p = 1 \end{cases}$$

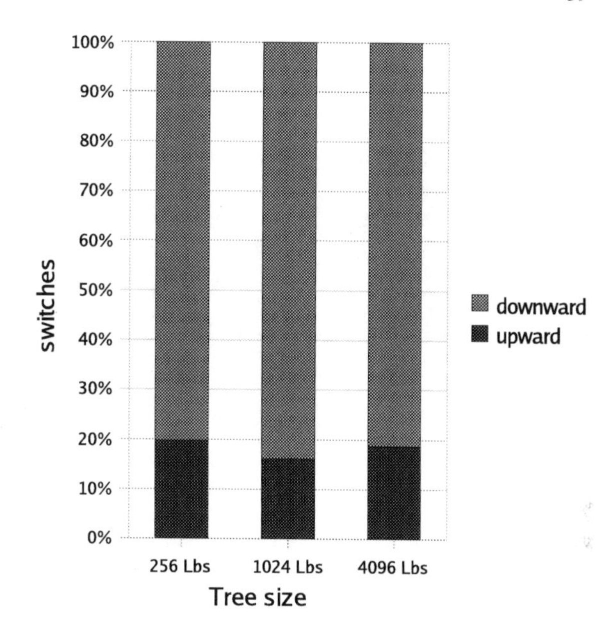

Fig. 3.8 Interconnect switches distribution

$$N_{switch}(LB) = \begin{cases} O(1) & \text{if } p < 1 \\ O\left(\log_k(N)\right) & \text{if } p = 1 \end{cases} \tag{3.2}$$

The cost of adding the upward can be compensated by reducing the architecture Rent's parameter. In addition we notice that the number of the upward network switches is smaller than the switches number in the downward network:

$$\frac{N_{switch}(down)}{N_{switch}(up)} = \frac{k^p c_{in} + k c_{out}}{k \times c_{out}}$$

With $p = 1, k = 4, c_{in} = 4$ and $c_{out} = 1$ this ratio is equal to 5. In Fig. 3.8, we show the distribution of interconnect resources between the upward and the downward networks for different Tree sizes (we include in/out pads connections).

Wiring Requirements

At each level ℓ of the hierarchy, every switching node has $n_{in}(\ell)$ inputs and $n_{out}(\ell)$ outputs. This makes the bisection width equal to $(c_{in} + c_{out})k^{\ell.p}$. Since $\forall \ell \in \{1, \ldots, \log_k(N)\}$ $k^{\ell.p} \leq N$, the bisection width is $O(N^p)$. For a 2-dimensional network layout this bisection width must cross the perimeter out of the subarray. Thus the perimeter of each subarray is $O(N^p)$. The areas of the subarray will be proportional to the square of its perimeter, making: $A_{subarray} \propto N^{2p}$. The required area per logic block (LB) based on wiring constraints, is therefore evaluated by:

$$A_{LB} \propto N^{2p-1}$$

In tree-based architecture, we can control bisection bandwidth in each level based on Rent's parameter ($p < 1$). Consequently, physical layout generation may be much optimized since wiring is no more dominant.

3.1.3 Comparison with Mesh Model

Concerning switches per logic block growth, it was established in [4] that in the Mesh architecture:

$$N_{switch}(LB) = O(N^{p-0.5}) \quad (3.3)$$

Equations (3.2) and (3.3) show that in the tree-based architecture, switches requirement grows more slowly than in common mesh-based architecture. These results are encouraging for constructing very large structures, especially when p is less than 1. But this does not mean that our Tree-based topology is more efficient than mesh-based architecture, since they do not have the same routability. The best way to check this point is through experimental work. Based on benchmark circuits implementation, we compare the resulting areas in the case of tree-based and mesh-based FPGA.

3.2 Architectures Exploration Environments

Since we are exploring two different architecture topologies, an effort is required to ensure transparency for comparison between two architectures. For this purpose, we have designed separate exploration environments for the two architectures. Some parts of two exploration environments are shared while rest of them are designed specifically to meet the needs of two architectures. Exploration of each architecture starts with its definition which is done through an appropriate architecture description mechanism. Once the architecture is defined, netlists are separately placed and routed on the two architectures. Although separate placement and routing tools are developed for two architectures, these tools are based on generic algorithms. These algorithms are adapted appropriately to the needs of the two architectures. Once the placement and routing of the netlists is performed, the area of the architectures is calculated which eventually leads to the termination of architecture exploration. Details of different steps that are involved in architecture exploration are explained in the following sections.

3.3 Architecture Description

3.3.1 *Architecture Description of Tree-Based Architecture*

In exploration environment of tree-based architecture, an architecture description mechanism is used to define different parameters of a tree-based architecture. The architecture description starts with the number of levels of the architecture. Then the level of I/O pads clusters and the number of I/Os per cluster are defined. Later the parameters of clusters located at all levels of the architecture are defined. These parameters include the arity and signal bandwidth of clusters that are located at that particular level. After that, architecture optimization is either set to be true or false. In case this parameter is set to be true, a binary search algorithm is applied to search the best signal bandwidth for the clusters that are located at different levels of hierarchy. Otherwise a fixed routing interconnect based on the initially defined cluster bandwidth values is built and no optimization is performed. Finally the parameters of CLBs are defined that include their level, area, inputs, outputs and details of their pins.

3.3.2 *Architecture Description of Mesh-Based Architecture*

The exploration environment of mesh-based architecture used in this book is based on the environment presented in [82]. In this environment, architecture description of mesh-based architecture starts with the definition of height and width of the slot-grid. Then Channel_Type of the architecture is defined which can be either a unidirectional mesh [77] or a bidirectional mesh [120] routing network. The channel width of the routing network is then either set to a constant value, or a binary search algorithm searches a minimum possible channel width. In case of unidirectional mesh, the channel width remains in multiples of 2. Finally the parameters of CLBs are defined which include a name, a size (number of slots occupied), area, number of inputs/outputs and the detail of their pins.

3.4 Software Flow

Once the architecture is defined, different netlists are placed and routed on the architecture using a software flow. However, before being placed and routed on an architecture, a netlist is required to pass through certain number of processes so that it might be converted from hardware description to a format that can be placed and routed on the FPGA architecture (.net format). Complete software flow illustrating these processes along with the placement and routing modules is shown in Fig. 3.9. As it can be seen from the figure that a certain part of software flow is shared by

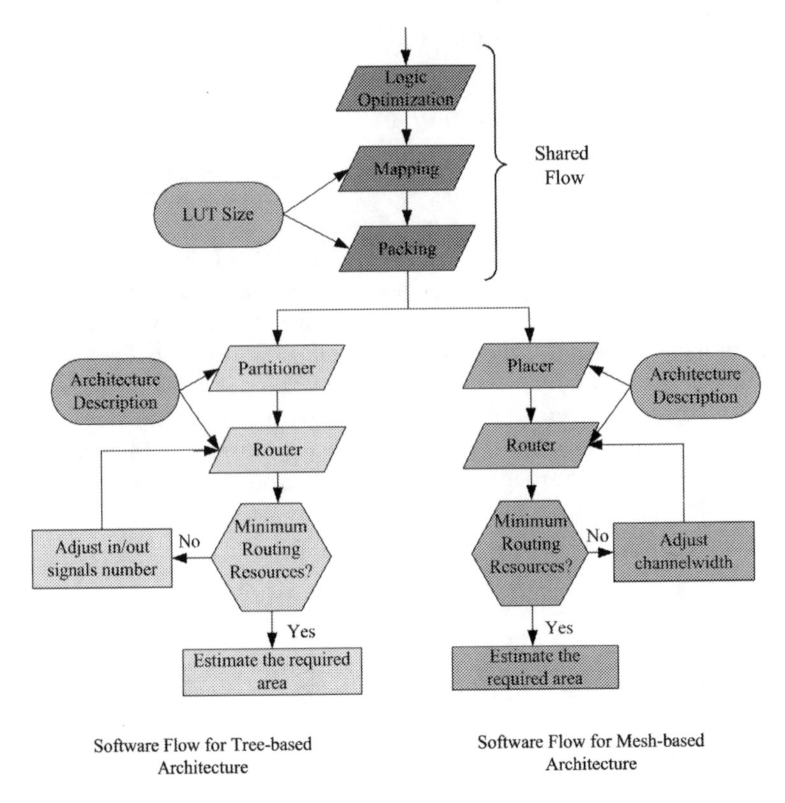

Fig. 3.9 Software flow

both mesh-based and tree-based architectures. This part involves the transformation of netlist from hardware description to .net format (i.e. synthesis of netlist) and once the netlist is converted into .net format, the two architectures place and route the netlist separately using their appropriate tools. A brief description of different tools involved in the software flow is given as follows:

3.4.1 Logic Optimization, Mapping and Packing

The input to the software flow is the hardware description of the netlist. First of all the netlist is synthesized/logically optimized using a tool called SIS [102]. This is a process in which circuit description is converted into gate level presentation. SIS is an open source tool and it can be replaced by any commercial synthesis tool.

After logic optimization, mapping of the netlist is performed using SIS. Mapping is a process that converts gate level representation into K-input LUT and flip-flops. This process takes LUT size as its input and converts logic expressions into given

LUT size netlist. We have used “Flow Map” [64] mapper for our experimentation which is included in SIS package. This mapper uses timing and area as its objectives and netlists produced using this mapper produce good results in terms of area and delay.

After mapping, packing is performed using T-VPACK [14] that packs registers together with K-input LUTs and converts the netlist into .net format. A netlist in .net format contains CLBs and I/O instances that are connected together using nets. The size of a CLB is defined as the number of LUTs contained in it and in this book this size is set to be 1 for both mesh-based and tree-based architectures. Once netlist is obtained in .net format, it is placed and routed separately on tree-based and mesh-based FPGA.

3.4.2 Software Flow for Tree-Based Architecture

3.4.2.1 Partitioning

In recent FPGA architectures, interconnect is organized in multiple hierarchical levels. Hierarchy becomes an interesting feature to improve density, to reduce run time effort (divide and conquer) and to consider local communication. In the case of a Tree-based interconnect we get multiple hierarchical levels. Levels number depends on the total number of CLBs and clusters size (arity). Basically if 2 signals are within the same hierarchy level, it does not really matter where they are within that hierarchy. Similarly, geometrically close cells incur greater delay to get to other locations outside their hierarchical boundary than to distant cells within their hierarchical boundary. Thus, unlike flat or island style device, a hierarchical architecture uses a natural placement algorithm based on recursive partitioning.

Multilevel hierarchical organization is considered in our CAD flow and netlists instances are partitioned between architecture clusters in the best possible way, reducing the desired objectives. There are two main partitioning approaches: bottom-up (clustering) and top-down. The choice between both approaches depends on levels number, clusters size, clusters number at each level and problem constraints. In [133], authors proposed to use hMetis [49] which uses an FM algorithm [47] based top-down partitioning approach for tree-based architecture. In fact top-down approaches based on FM refinement heuristics are efficient when we target a small number of clusters (parts) of important size (balance constraint). To investigate partitioning approaches, we used a multilevel hypergraph data structure called *Mangrove*. It provides a development framework for efficient modeling of hypergraph nested partitions. It offers a compact C++ data structure and a high level API. As illustrated in Fig. 3.10, this structure is organized as follows:

- *ClusteringHierarchy*: holds a vector of nested partitions called *Clustering Level*, and refers to a unique enclosing cluster *TopLevelCluster*,

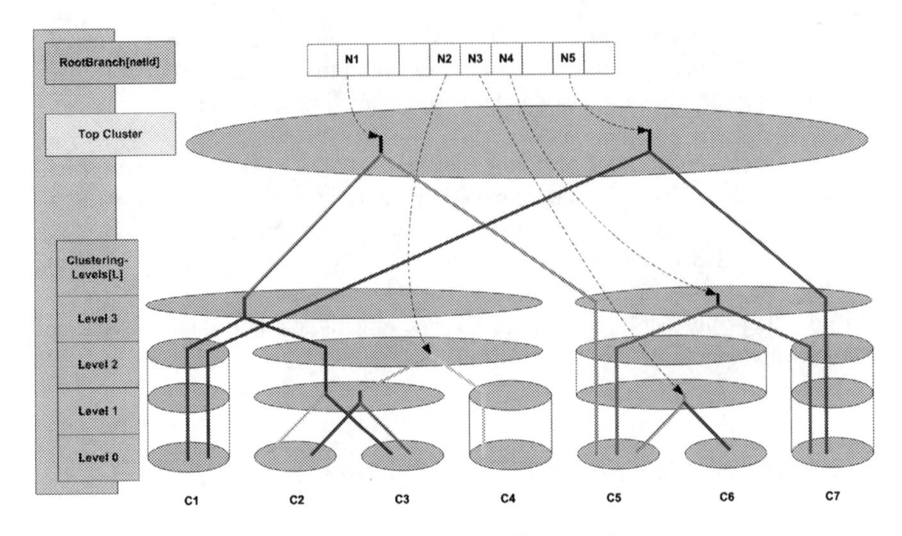

Fig. 3.10 Mangrove data structure: multilevel clustered hypergraph

- $ClusteringLevel$: corresponds to the set of clusters at the partitioning at a given level. A $clusteringLevel$ corresponds to an hypergraph where nodes are clusters located at this level,
- $Cluster$: Aggregates sub-clusters belonging to a lower $ClusteringLevel$ (unless leaf one). A $Cluster$ may cross multiple levels and has $UpperLevel$ and $LowerLevel$ identifiers,
- Net: presents a tree of branches,
- $Branch$: represents the net (signal) crossing point of a cluster boundary. Branch bifurcates within a cluster if the net crosses at least 2 sub-clusters.

Since in Mangrove a $clusteringLevel$ can be added at any level, this structure can be used in different partitioning approaches: Bottom-up and top-down. The combination of both approaches leads to an efficient multilevel partitioner where first multilevel bottom-up coarsening is run and then top-down multilevel refinement is applied. In Fig. 3.11, we show different steps of recursive netlist partitioning based on a multilevel approach. The netlist is first partitioned into 2 parts (first level) and then instances inside each part are partitioned into 2 fractions. In each partitioning phase we apply a multilevel coarsening followed by a multilevel refinement. Finally, we obtain the partitioning result corresponding to each level. The final result describes how instances are distributed between clusters of the Tree-based topology. Recursive partitioning is also interesting to reduce run time since it allows to avoid applying FM heuristics directly on a large number of parts, which can dramatically increase partitioning run time according to [75].

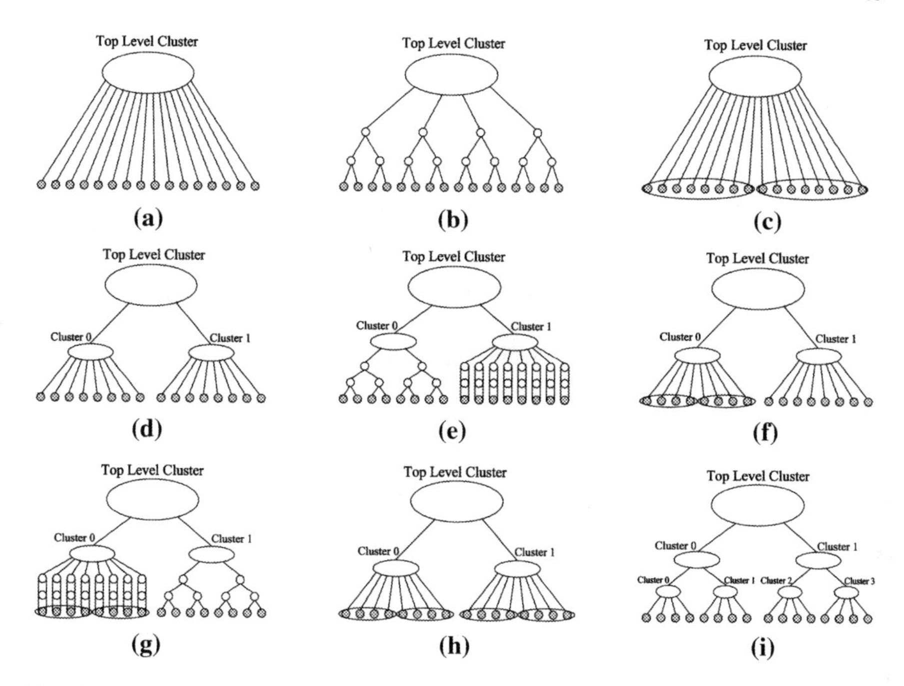

Fig. 3.11 2-levels recursive bi-partitioning steps. **a** Leaves at level 0 hypograph partitioning, **b** coarsened hypergraph, **c** bi-partitioned hypergraph, **d** clustering according to partitioning, **e** restricted coarsening in cluster 0, **f** bi-partitioning in sub hypergraph (cluster 0), **g** restricted coarsening in cluster 1, **h** bi-partitioning in sub hypergraph (cluster 1), **i** clustering according to partitioning

3.4.2.2 Routing

Once partitioning is done, placement file is generated that contains positions of different blocks on the architecture. This placement file along with netlist file is then passed to another software module called router which is responsible for the routing of netlist. In order to route all the nets of netlist, routing graph is first constructed that contains the details about all the routing resources of the architecture. The construction of routing graph is mainly dependant upon the cluster bandwidth information which is extracted from the architecture description file. Once the routing graph is constructed, routing resources of the architecture are later assigned to respective blocks of the netlist that are placed on the architecture. These routing resources are modeled as directed graph abstraction $G(V, E)$ where the set of vertices V represents the in/out pins of different blocks and the routing wires in the interconnect structure and an edge E between two vertices, represents a potential connection between the two vertices. Router is based on PathFinder [80] routing algorithm that uses an iterative, negotiation-based approach to successfully route all nets in a netlist.

3.4.3 Software Flow for Mesh-Based Architecture

3.4.3.1 Placement

For mesh-based architecture, the netlist obtained in the .net format is placed using a placement algorithm. The placement algorithm determines the position of block instances of a netlist on their respective block types on FPGA architecture. The main goal is to place connected instances near to each other so that minimum routing resources are required to route their connections. The placer uses simulated annealing algorithm [37, 105] to achieve a placement having minimum sum of half-perimeters of the bounding boxes of all the nets.

3.4.3.2 Routing

After the placement of netlist on the FPGA architecture, the exploration environment constructs routing graph for the architecture. Few architecture description parameters required for the construction of routing graph are taken from architecture description parameters. These parameters mainly include the type of routing network (unidirectional or bidirectional), channel width, I/O rate, block types and their pin positions on the block. After the construction of routing graph, the PathFinder routing algorithm [80] is used to route netlists on the routing architecture.

3.4.4 Timing Analysis

Timing analysis evaluates performances of a circuit implemented on an FPGA (mesh or tree) in terms of functional speed. Thus, once an application is completely placed and routed we estimate the minimum feasible clock separately for mesh and tree-based architectures. To achieve timing analysis we need 2 different graphs:

- Routed graph: Describes the way netlist instances are routed using architecture resources. This graph allows to evaluate routing delays between netlist instances connections. A path connecting two instances crosses several wires and switches. The connection delay is equal to the sum of resources delays.
- Timing graph: It is a direct acyclic graph generated from the netlist hypergraph. Nodes correspond to instances pins and edges to connections. Based on the resulting routed graph, each edge is labeled with the corresponding routed connection delay. The minimum required clock period is determined via a breadth-first traversal applied on this graph.

Only the routed graph is architecture dependent. Timing graph generation and critical path extraction depend only on netlist to implement.

3.4.5 Area and Delay Models

Once the netlists are placed and routed on the architecture, area and performance of the architecture is calculated using respective area and delay models. These models are generic in nature and they are applicable to both mesh-based and tree-based architectures. This section describes the generic area and delay models that we have used for both architectures.

3.4.5.1 Area Model

The area model is used to compute the areas of two architectures under consideration and these architectures are then compared using the area values calculated through their respective models. As mentioned in [22], discussions with FPGA vendors have revealed that transistor area, and not wiring density, is the area limiting factor. The use of directional wires in Virtex-I also suggests that routing area is transistor-dominant and must be reduced. As it was explained by DeHon [5], large area of switches compared to wires is one of the key reasons why we have to care about the number of switches required by a network. If the wire pitch is 5–8λ, the area of a wire crossing is 25–64λ^2. The area of static memory cell used to configure a switch is roughly 1,200λ^2. A switch transistor size is 2,500λ^2. In this case the ratio switches area/wires area can reach the value of 40. This ratio increases if we want more than just a pass gate for the switch. We may want to rebuffer the switch or even add a register to it. Such switch can easily be 5–10Kλ^2. The large area ratio means that we definitely need to take much care about switch count in the interconnect.

Area of any FPGA architecture can be basically divided into two parts: logic area and routing area. Logic area is a small part of total area and it comprises of the area of logic cells (i.e. CLBs) of the architecture. Routing area, on the other hand, comprises of the area of switching cells used by the routing network of the architecture and it can take up to 90% of the total area of the architecture. Routing area of the architecture includes area of configuration memory, multiplexors and buffers etc. An example showing how these switching cells are combined to construct a routing interconnect is shown in Fig. 3.12. Area of SRAMs, multiplexors, buffers and Flip-Flops is taken from a symbolic standard cell library (SXLIB [9]) which works on unit Lambda(λ). Area of different cells used for the area calculation is shown in Table 3.1.

3.4.5.2 Delay Model

The delay through the routing network may easily be dominant in a programmable technology. Care is required to minimize interconnect delays. The 2 following factors are significant in this respect:

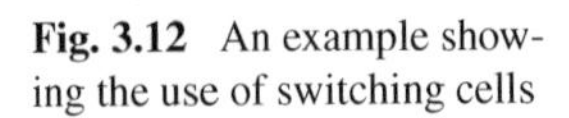

Fig. 3.12 An example showing the use of switching cells

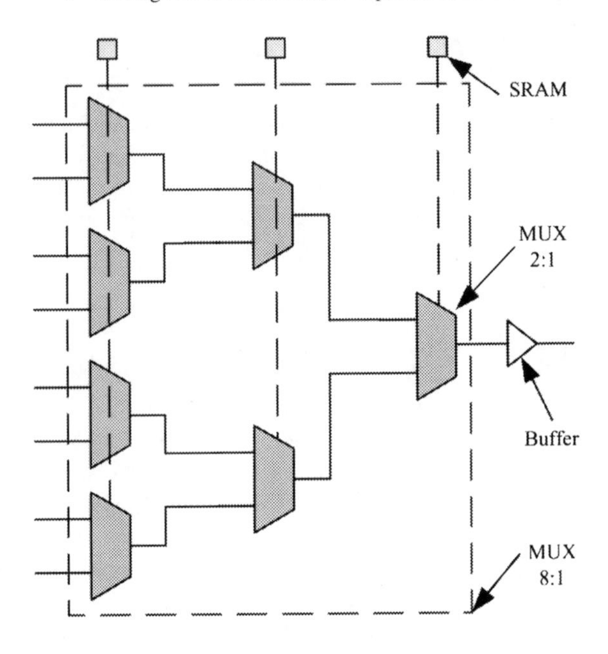

Table 3.1 Area of different cells

Block name	Inputs	Outputs	Block size (λ^2)
clb	4	1	58,500
sram	–	–	1,500
buffer	1	1	1,000
flip-flop	1	1	4,500
mux 2:1	2	1	1,750

- Wires delay: Delay on a wire is proportional to distance and capacitive loading (fanout). This makes interconnect delay roughly proportional to distance run. Consequently short signals runs are faster than long signals runs.
- Switches delay: Each programmable switche in a path (crossbar, multiplexer) adds delay. This delay is generally much larger than the propagation or fanout delay. Consequently, one generally wants to minimize the number of switch elements in a path, even if this means using some longer signals runs.

Wire length and switches delays depend respectively on physical layout and cells library.

Table 3.2 Description of circuits used in experiments

Index	Circuit name	No of inputs	No of outputs	No of 4-input LUTs
1	pdc	16	40	3,832
2	ex5p	8	63	982
3	spla	16	46	3,045
4	apex4	9	19	1,089
5	frisc	20	116	2,841
6	apex2	38	3	1,522
7	seq	41	35	1,455
8	misex3	14	14	1,198
9	elliptic	131	114	2,712
10	alu4	14	8	1,242
11	des	256	245	1,506
12	s298	4	6	1,091
13	bigkey	229	197	1,147
14	diffeq	64	39	1,161
15	dsip	229	197	1,145
16	tseng	52	122	953

3.5 Experimentation and Analysis

Exploration environments described in the previous section are used to place and route different netlists on the two architectures and results are later compared to evaluate them. We have used 16 largest MCNC benchmarks for our experimentation. Details of these benchmarks are shown in Table 3.2. Name and I/Os of the circuits under consideration are shown in first three columns of the table whereas the size of each benchmark in terms of the number of 4 input LUTs used by it is shown in the last column of the table.

3.5.1 Architectures Optimization Approaches

The benchmarks shown in Table 3.2 are individually placed and routed on the two architectures under consideration. Specifically developed optimization approaches are used for both architectures to get the optimized area and delay results for both architectures. An overview of these optimization approaches is given below.

3.5.1.1 Tree-Based Architecture Optimization

Optimization of the tree-based architecture is dependant upon the information given in the architecture description file. If the optimization flag is false, the routing of the

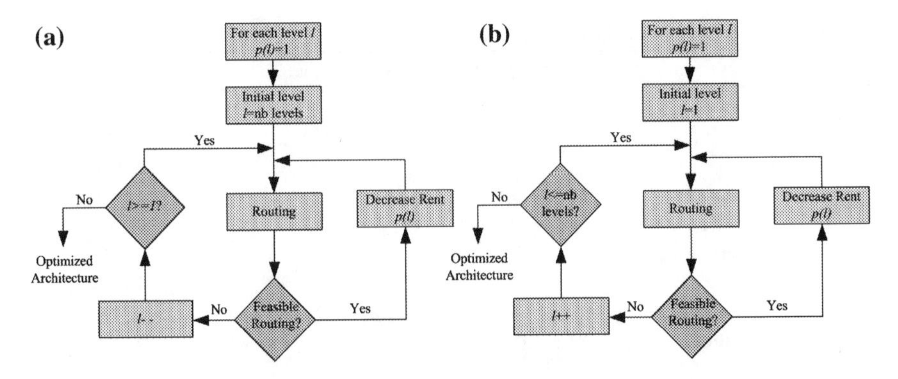

Fig. 3.13 Tree-based architecture optimization flow. **a** Top-down architecture optimization, **b** bottom-up architecture optimization

netlist is performed with given signal bandwidth and the exploration of the architecture is terminated after the calculation of area. However, if the optimization flag is true, a binary search algorithm is used to find the minimum signal bandwidth required to route the netlist. As explained in Sect. 3.1.2 the optimization of a tree-based architecture is dependant upon N and Rent's parameter p, the optimization algorithm that we employ is used to optimize the value of p for each level of the tree-based architecture. We apply binary search algorithm on each level to determine the minimum I/O bandwidth required at that particular level and this process continues until all the levels of the architecture are optimized. The value of p is then averaged across all the levels to determine the p for architecture. Although the clusters situated at different levels of hierarchy may have different values of p, clusters located at same level have same value of p. Based on the level order, we have explored three different types of optimization approaches for tree-based FPGA architecture. A brief overview of these approaches is described as follows:

1. Top-down approach: As shown in Fig. 3.13a, we start by optimizing the top level down to the lowest one. At each level we apply binary search to determine the smallest input/output signals number allowing to route the benchmark circuit.
2. Bottom-up approach: As shown in Fig. 3.13b, we start by optimizing the lowest level up to the highest one. At each level we apply binary search to determine the smallest input/output signals number allowing to route the benchmark circuit.
3. Random approach: All levels are optimized simultaneously. We choose a level randomly, we decrease its input/output signals number, depending on the previous result obtained in this level; then we move to an other level. In this way we move randomly from a level to another until all levels are optimized.

The 3 approaches have the same objective and aim at reducing clusters signals bandwidth for every level. The difference is the order in which levels are processed. In Table 3.3, we show architecture Rent's parameter (in each level) obtained with each technique. The first column of the table shows Rent's parameters, at each level, obtained after circuits partitioning. Results correspond to averages of all 16 circuits.

Table 3.3 Levels Rent's rule parameters

Level	Circuits partitioning	Architecture top-down	Architecture bottom-up	Architecture random
1	0.64	0.98	0.79	0.88
2	0.55	0.88	0.74	0.79
3	0.50	0.80	0.77	0.76
4	0.49	0.75	0.86	0.73
5	0.45	0.59	0.87	0.7

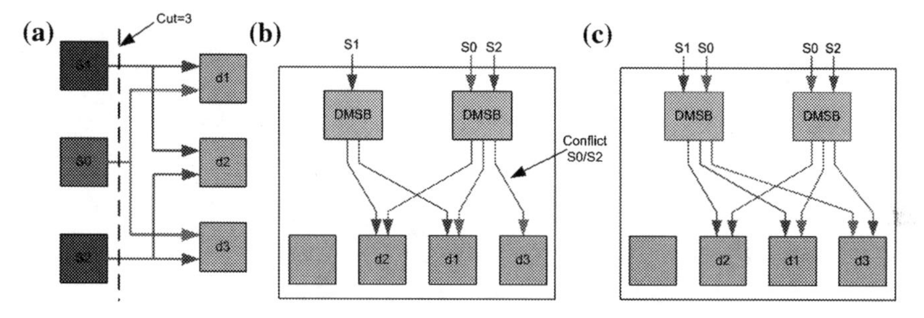

Fig. 3.14 A netlist routing example. **a** Partitioned netlist, **b** routed netlist with conflict, **c** routed netlist with no conflict

We notice that in all cases, architecture Rent's parameters are larger than partitioned circuits Rent's parameters. This is due to the depopulated switch boxes topology. In fact, to solve routing conflicts, a signal may enter from 2 different DMSB to reach 2 different destinations located at the same cluster. In Fig. 3.14 we show an example of a partitioned netlist to place and route on an architecture with LBs inputs number equal to 2 (2 DMSBs in each cluster located at level 1) and clusters size equal to 4. As shown in Fig. 3.14, if each signal enters from only one DMSB, we cannot solve conflicts. To deal with such problem we propose to enter the signal driven by S0 from two different DMSBs. Thus, the resulting architecture cluster degree is equal to 4, whereas the corresponding part degree is equal to 3 (number of crossing signals).

In Fig. 3.15, we show the average overhead between partitioning and architecture Rent's parameters with each optimizing approach. We notice that in the case of the top-down (bottom-up) approach, overhead increases when we go down (up) in the Tree. This was expected since the top-down (bottom-up) approach first optimizes high (low) levels. With the random approach, we notice that levels overheads are balanced.

We compared the resulting architectures (with the 3 approaches) in terms of area and speed performance. Average results are shown in Table 3.4. We notice that with the random approach we obtain the smallest area (22% less than top-down and 8% less than bottom-up). This means that optimizing levels simultaneously allows avoiding local minima and obtaining a balanced congestion distribution over levels. The bottom-up approach provides a smaller area than the top-down one. Nevertheless,

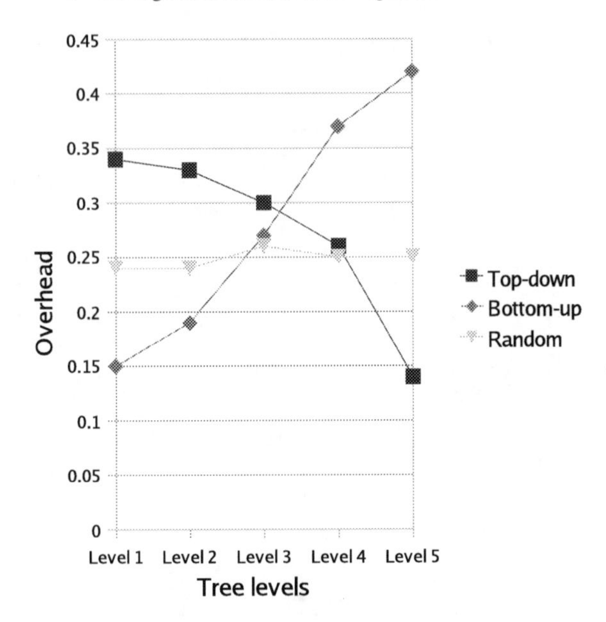

Fig. 3.15 Overhead between architecture and partitioned netlist Rent's parameters (16 benchmark avg.)

Table 3.4 Area and performance comparison between various optimizing approaches

Optimizing approach	Area (λ^2) $\times 10^6$	Critical path switches
Top-down	1,498	98
Bottom-up	1,326	106
Random	1,221	101

it is penalizing in terms of critical path switches number (8% more switches than top-down approach). In fact starting by optimizing low levels means that local routing resources are intensively reduced and signals are routed with resources located at higher levels. Consequently, signals routing uses more switches in series.

To reduce the gap between circuit and architecture Rent's parameters, we must improve the partitioning tool and especially the objective function in order to reduce congestion and resources (clusters inputs) required to route signals.

3.5.1.2 Mesh-Based Architecture Optimization

Like tree-based architecture, the optimization of mesh-based architecture is also dependant on the information given in the architecture description file. If the binary search flag is false, routing of the netlist is performed using a given value of channel width and the experimentation is terminated with the area calculation. However if the binary search flag is true, routing graph is constructed for varying channel widths;

routing is tried for each channel width until a minimum channel width is found. This optimization approach is similar vpr-based optimization approach used in [22].

3.5.2 Effect of LUT and Arity Size on Tree-Based FPGA Architecture

Before we start with the comparison between two FPGAs under consideration, effect of LUT (K) and arity (N) size is first explored for tree-based FPGA architecture. Many studies in the past several years have been carried out to see the effect of LUT and cluster size on the density of mesh-based FPGA architecture. The work in [8] compiles a very detailed study regarding the effect of LUT and cluster size on the density and performance of FPGA architecture. In [8] the authors have shown that LUTs with sizes 4 to 6 and clusters with sizes 3 to 10 give the most efficient results in terms of area-delay product for an FPGA. The work in [71] demonstrated that LUT size of 4 is most area efficient in a non clustered context. But all the work previously done in this context focusses on the mesh-based architecture and no prior work has been done yet for tree-based architectures. In this work, we first start our experimentation by exploring the effect of LUT and arity size on a tree-based architecture. This exploration is significant in the sense that an appropriate combination of K and N plays an important role in the overall efficiency of the architecture. In order to perform the exploration we use 16 MCNC [108] benchmarks shown in Table 3.2. These benchmarks are generated with different LUT (K) and arity sizes (N) and then they are placed and routed on the tree-based architecture using the flow described in Sect. 3.4. For these benchmarks, LUT size is varied from 3 to 7 while arity size is varied from 4 to 8.

Effect of LUT (K) and arity size (N) on area and performance of tree-based FPGA architecture is shown in Figs. 3.16 and 3.17 respectively. It can be seen from Fig. 3.16 that for almost all arity sizes, there is a reduction in total average area from LUT-3 to LUT-4. However, as the LUT size is increased beyond LUT-4, the total area of the architecture increases. It can be concluded from this figure that $K=4$ with $N=4$ gives best overall area results.

The second key metric is the effect of K and N on the performance of tree-based architecture. Since we do not have accurate wire lengths, only results regarding number of switches crossed by critical path are presented here. The extraction of number of switches crossed by critical path is explained in Sect. 3.4.4. Figure 3.17 shows the effect of varying K and N on critical path performance of the architecture. Results presented in this figure correspond to average number of switches crossed by 16 benchmarks under consideration. It is clear from this figure that an increase in K or N decreases the average number of switches that are crossed by critical path. This is because of the fact that an increase in K or N decreases the architecture size and there are smaller number of switches on the critical path. However, an increase in K or N increases the size of these switches; hence resulting in an increase in the

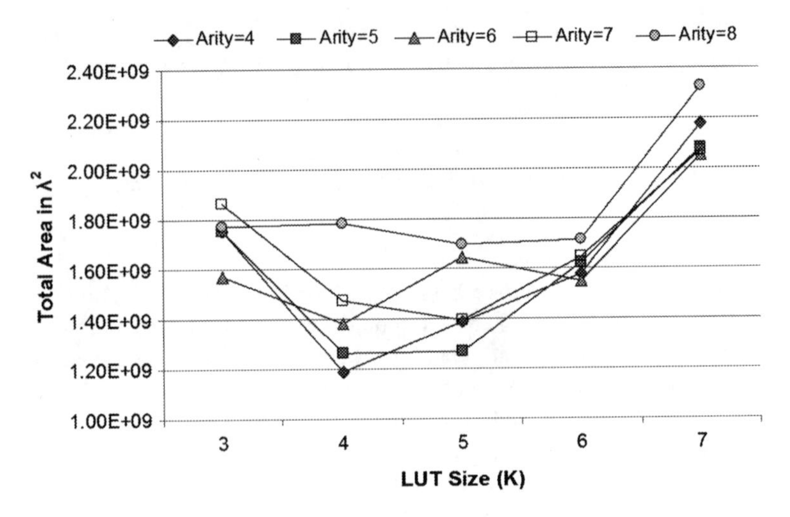

Fig. 3.16 Effect of LUT and arity size on total area of tree-based FPGA architecture

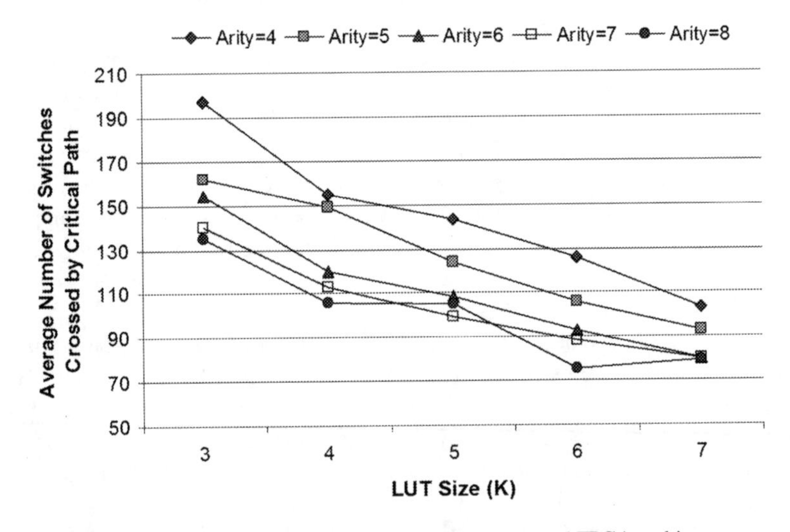

Fig. 3.17 Effect of LUT and arity size on critical path of tree-based FPGA architecture

intrinsic delay of the switch. So from this figure we can conclude that with a increase in K and N there will be decline in the critical path delay provided that the increase in internal delay of switch due to its increased size does not overshadow the reduction in number of switches. Further details regarding the effect of K and N on tree-based architecture can be found in [46] where we have come up with the conclusion that overall K = 4 with N = 4 provide the best overall results and this is the combination that will be used for tree-based FPGA architecture in this work.

Table 3.5 Architecture characteristics

Circuit name	Mesh-based architecture			Tree-based architecture		
	Architecture N×N	Occupancy (%)	Channel width	Architecture levels	Occupancy (%)	Rent's p
pdc	64 × 64	93	20	4 × 4 × 4 × 4 × 4 × 4	93	0.98
ex5p	32 × 32	95	16	4 × 4 × 4 × 4 × 4	96	1.00
spla	57 × 57	93	18	4 × 4 × 4 × 4 × 4 × 4	74	0.90
apex4	34 × 34	94	16	4 × 4 × 4 × 4 × 4 × 2	53	0.83
frisc	55 × 55	94	14	4 × 4 × 4 × 4 × 4 × 4	69	0.86
apex2	40 × 40	95	14	4 × 4 × 4 × 4 × 4 × 2	74	0.91
seq	39 × 39	96	16	4 × 4 × 4 × 4 × 4 × 2	71	0.89
misex3	36 × 36	92	14	4 × 4 × 4 × 4 × 4 × 2	58	0.84
elliptic	53 × 53	97	12	4 × 4 × 4 × 4 × 4 × 4	66	0.80
alu4	36 × 36	96	14	4 × 4 × 4 × 4 × 4 × 2	60	0.83
des	40 × 40	94	10	4 × 4 × 4 × 4 × 4 × 2	73	0.91
s298	34 × 34	94	12	4 × 4 × 4 × 4 × 4 × 2	53	0.77
bigkey	35 × 35	93	8	4 × 4 × 4 × 4 × 4 × 2	56	0.74
diffeq	35 × 35	95	10	4 × 4 × 4 × 4 × 4 × 2	56	0.72
dsip	35 × 35	93	8	4 × 4 × 4 × 4 × 4 × 2	56	0.74
tseng	31 × 31	99	8	4 × 4 × 4 × 4 × 4	93	0.88
Average	–	94	14	–	68	0.85

3.5.3 Comparison Between Homogeneous Mesh and Tree-Based FPGAs

In [132], a comparison between mesh-based and tree-based FPGAs is performed and it is shown that on average tree-based FPGA is 56% better than mesh-based FPGA in terms of area. However, the reference mesh-based architecture used in [132] is a bidirectional FPGA architecture and in [77], authors have proposed to replace the bidirectional interconnect with the unidirectional interconnect as the later gives better results compared to the former. So in this chapter we have changed the reference mesh-based architecture from bidirectional to unidirectional and we have re-evaluated the tree-based architecture. For tree-based architecture the LUT size is set to be 4 while arity size is set to be 4 too as it gives best overall results for it and for mesh-based architecture the LUT size is also set to be 4 and CLB size is set to be 1 for both architectures.

Experimental results of the two architectures are shown in Tables 3.5 and 3.6 respectively. Experiments are performed for individual netlists where an appropriate architecture is defined for each netlist and architecture is optimized using the optimization algorithm described in previous section. Although individual optimization approach has been, at times, controversial because most engineers think that FPGA is a fixed device and it does not vary in response to individual circuits that are being mapped on it. However, this more refine approach is usually used as it is necessary

Table 3.6 Comparison results between mesh-based and tree-based architectures

Circuit name	Mesh-based architecture			Tree-based architecture			Gain		
	Area $\times 10^6\lambda^2$	SRAMs $\times 10^3$	MUXs $\times 10^3$	Area $\times 10^6\lambda^2$	SRAMs $\times 10^3$	MUXs $\times 10^3$	Area (%)	SRAMs (%)	MUXs (%)
pdc	2,756	425	810	1,344	289	404	51	32	50
ex5p	567	84	162	296	62	86	47	25	46
spla	1,994	304	577	1,078	211	305	46	30	47
apex4	639	95	183	441	83	114	30	12	37
frisc	1,483	221	415	965	185	264	34	16	36
apex2	787	118	220	525	106	146	33	9	33
seq	839	124	240	480	93	128	42	24	46
misex3	639	95	179	417	77	105	34	19	40
elliptic	1,208	183	329	812	154	201	32	15	38
alu4	639	95	179	433	80	113	32	15	36
des	572	93	158	523	85	144	8	8	9
s298	500	76	136	378	66	92	24	12	32
bigkey	364	56	96	348	56	81	4	0.54	16
diffeq	432	70	119	330	56	74	23	20	38
dsip	364	56	96	347	56	81	4	0.64	15
tseng	281	43	74	242	41	54	13	5	26
Average	879	134	248	559	106	150	29	15	34

to evaluate the quality of an architecture in a more precise manner [7]. Different architectural parameters for the two architectures (architecture size, occupancy and signal bandwidth) are shown in Table 3.5 where individual architecture is defined and optimized for each of the netlist under consideration. It can be seen from the table that, compared to the occupancy of mesh-based architecture, tree-based architecture has a smaller average occupancy. This smaller occupancy of tree-based architecture is due to its hierarchical nature and compared to mesh-based architecture the logic resources of the tree-based architecture are under utilized. However, poor logic utilization is remedied by spreading the congestion of interconnect resources (congestion spreading effect is explained below) which eventually leads to better area results compared to mesh-based architecture.

Area results of the two architectures are shown in Table 3.6. It can be seen from the table that tree-based architecture gives better area results for all the netlists and on average it gives 29% area gain compared to unidirectional mesh-based architecture. One of the reasons of this area gain is the ability of tree-based architecture to simultaneously control the logic occupancy and interconnect population. It can be seen from Table 3.5 that generally the netlists with higher occupancy have a higher Rent's p (e.g. netlist named ex5p) and the netlists with smaller occupancy have a smaller Rent's p. This confirms that we can balance the interconnect and logic utilization by decreasing the logic occupancy and spreading the congestion. In fact, for tree-based architecture, we use a high-interconnect/low-logic utilization which is in

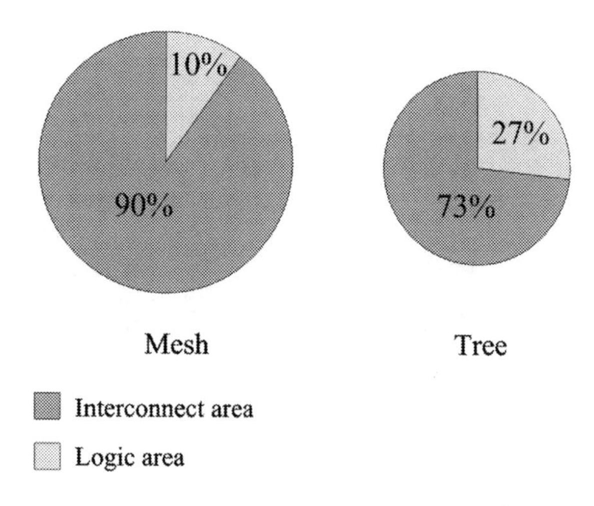

Fig. 3.18 Area distribution between interconnect and logic area for mesh-based and tree-based architectures

contrast to the mesh architecture's high-logic utilization approach. It can be seen from Fig. 3.18, unlike mesh-based architecture where interconnect occupies 90% of total area, in tree-based architecture interconnect occupies 73% of total area. This is because of the fact that in tree-based architecture, lower logic occupancy leads to higher interconnect depopulation which ultimately leads to better area results for tree-based architecture.

3.6 FPGA Hardware Generation

This section presents an automated method of generating hardware description of mesh-based and tree-based FPGA architectures. For both mesh-based and tree-based architectures, the FPGA hardware generator is integrated with their exploration environment. By doing so, all FPGA architectural parameters that are supported by the exploration environment are automatically supported by the VHDL model generator. Different size of FPGAs, with varying signal bandwidths, different variety of blocks, modification in connection patterns are automatically supported by this VHDL model generator. The VHDL model is passed to Cadence Encounter to generate layout of FPGA for 130 nm 6-metal layer CMOS process of ST Micro-Electronics. Different steps involved in the FPGA hardware generation are described in the sections that follow where the hardware generation of both mesh-based and tree-based architectures is detailed. Although this chapter considers only the generation of VHDL model of two FPGA architectures, an approximate layout scheme is taken into account to generate efficient VHDL model for two architectures. This scheme considers the efficient distribution of logic and routing resources of the architecture and it tries to distribute them in a uniform manner; hence eventually leading towards a less

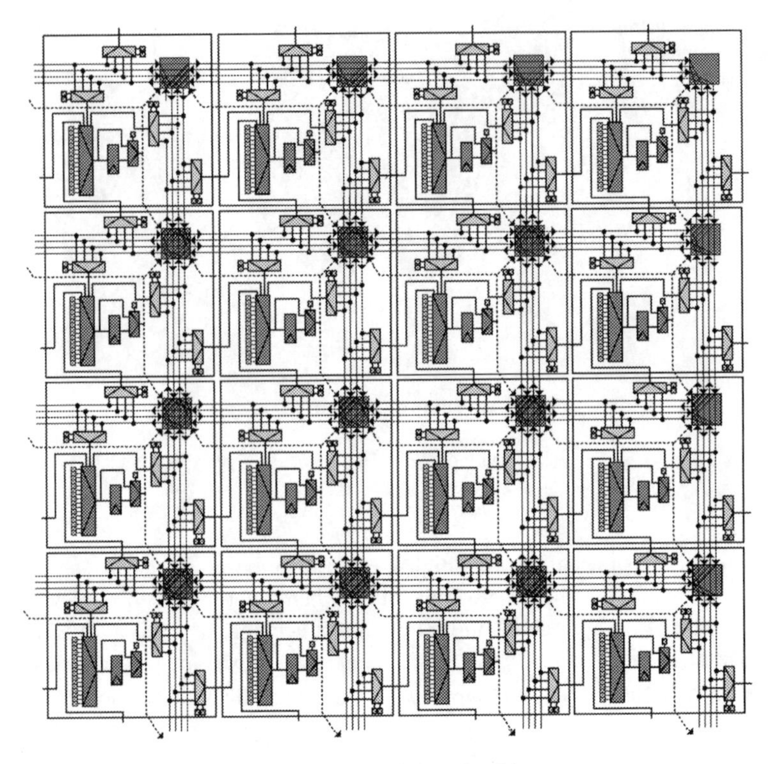

Fig. 3.19 Detailed interconnect of 16 tiles of mesh-based architecture

congested architecture. An overview of the two architectures is already presented in Sect. 3.1.

As discussed earlier, mesh-based architecture uses a unidirectional interconnect and the detailed interconnect of different tiles of mesh-based architecture is shown in Fig. 3.19. In this figure the detailed interconnect of 16 tiles is shown that are arranged in a 4x4 manner. As it can be seen from the figure that each tile contains switch box, a connection box, a CLB and routing wires on top and right side of CLB. These tiles can be further replicated to build larger architectures. In order to generate an optimized VHDL model of the architecture, these tiles are numbered in an appropriate manner and the logic and routing resources covered by each tile are also numbered. These numbers are later used while generating the VHDL model of the architecture.

For tree-based architecture, the approximate layout scheme needs an arrangement of logic and routing resources and it can be explained with the help of Fig. 3.20. In this figure, second level cluster of a tree-base architecture is shown where this cluster contains four level 1 clusters and each level 1 cluster in turn contains 4 CLBs which makes it a part of arity-4 tree-based architecture. Contrary to Fig. 3.6, in this figure switches and wires are depicted in different colors and for clarity only a small portion of the total interconnect is shown here. This is done to differentiate between the switches and wires of downward and upward interconnect of different levels. It

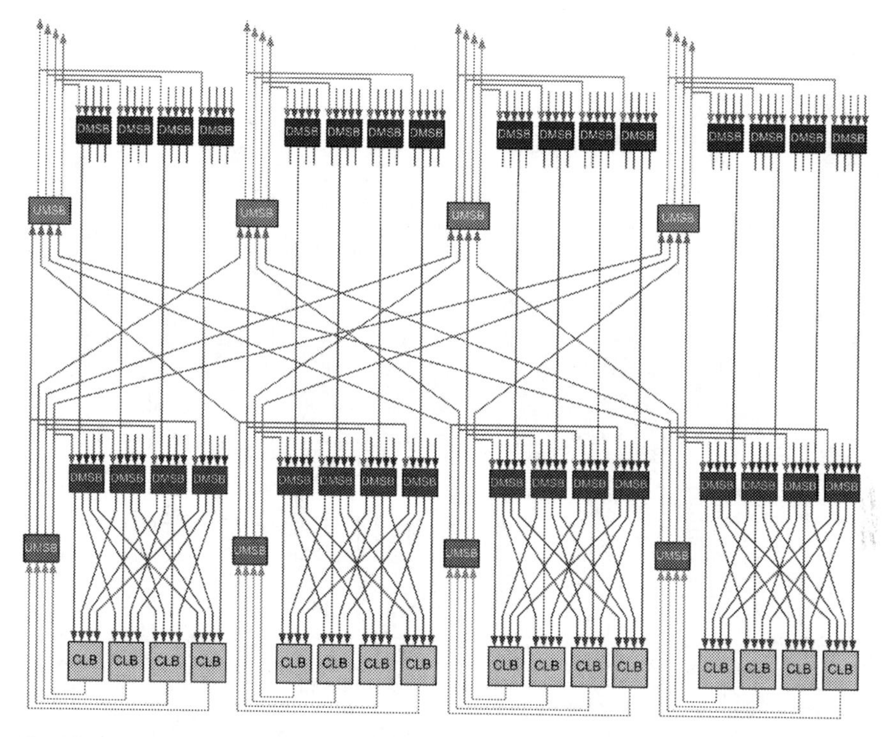

Fig. 3.20 Sparse interconnect of a level 2 cluster of a tree-based architecture

can be seen from this figure that every DMSB of level 1 cluster and its wires are shown in blue and every UMSB of level 1 and its wires are shown in red. Similarly for level 2 cluster the color is green for DMSBs and orange for UMSBs.

The approximate layout scheme of this cluster is shown in Fig. 3.21. This is a detailed but topologically equivalent scheme of the cluster shown in Fig. 3.20. In this figure switch blocks are broken down into small switches where each switch is placed horizontally or vertically in front of its successor. The division of a switch block into smaller switches is explained in Fig. 3.22 where a full cross bar switch has 5 inputs and four outputs and it is divided into programmable switches (multiplexors). Every programmable switch is composed of a group of switches and these groups are either placed in the same row or same column as their successor. CLBs of level 1 cluster of Fig. 3.20 are arranged in rows and the interconnect of two levels is interwoven uniformly to build a regular structure based on tiles. Each tile in Fig. 3.21 contains a CLB, a set of level 1 switches and a set of level 2 switches. In order to vary the arity of the architecture, we vary the number of CLBs in a row and in order to vary the number of inputs and outputs in a cluster level, we vary the number of switches in the tile. This corresponds to change in multiplexor size in level 1 or level 2. Thus scalability is taken care of in terms of arity and number of I/Os. It can be seen from the figure that all tiles are equivalent in terms of logic and switches distribution. Similarly all

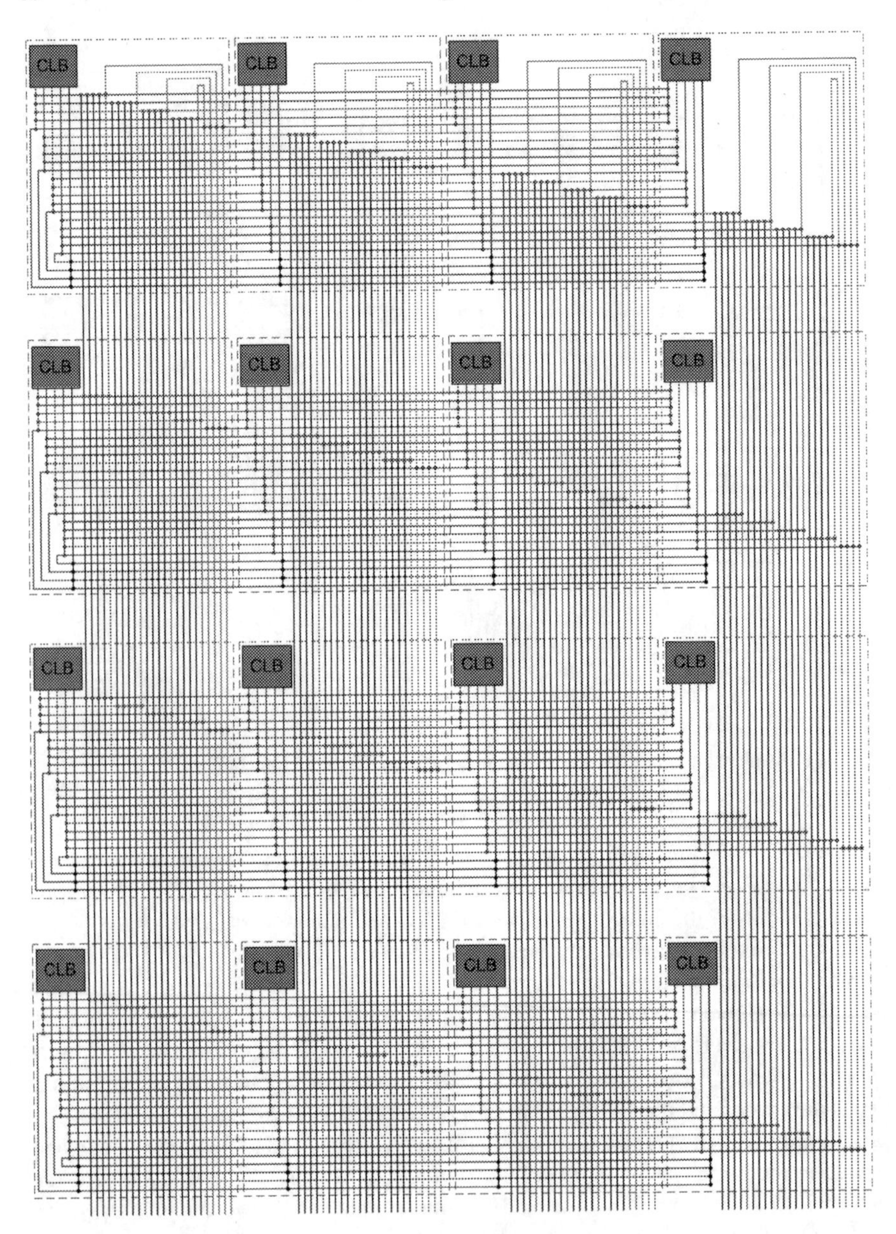

Fig. 3.21 Floor-plan of level 2 cluster of tree-based FPGA architecture

tiles of a column are equivalent. Although tiles of same row are different in routing topology, they are still equivalent in terms of number of switches. To generate the VHDL model for larger tree-based architectures, same technique of interweaving is applied as it gives flexibility in terms of arity and number of inputs and outputs of a

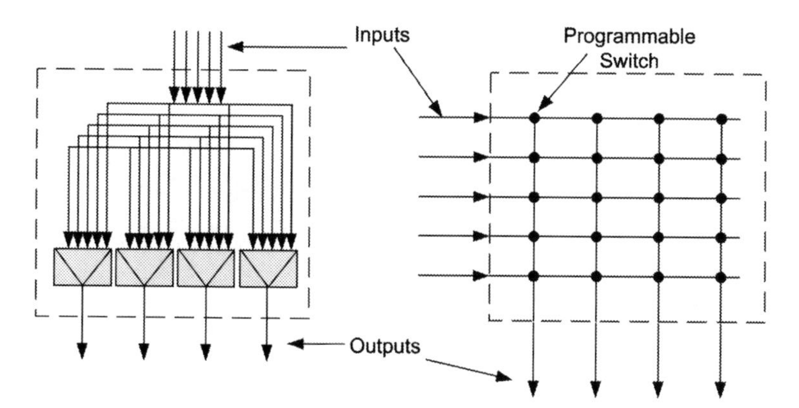

Fig. 3.22 Detailed topology of a switch block

cluster. However, it is important to mention here that it is just an approximate scheme and it does not represent the custom layout of tree-based architecture.

3.6.1 FPGA Generation Flow

Figure 3.23 shows the generalized flow that is used to generate the FPGA hardware in an automated manner. This flow is applicable to both mesh-based and tree-based architectures except that the two architectures use their respective architecture description mechanisms and exploration environments to generate the hardware. As it can be seen from the figure that the flow takes three parameters as its input:

1. Database of blocks contains the blocks definition of all the blocks that are supported by the FPGA architecture.
2. Architecture description file contains different architecture parameters using which the architecture is built. Details of the architecture description and block definition mechanism in mesh-based and tree-based architectures is given in Sect. 3.3.
3. Third parameter of the flow is the netlist that contains the blocks that are defined using block definition database and this netlist is then placed and routed on the architecture which is constructed using the parameters of architecture description file.

FPGA architecture exploration environment uses the architecture description file to construct the architecture and later the netlist is efficiently placed and routed on this architecture. Once the netlist is placed and routed on the architecture, exploration environment generates results like the floor-planning and routing graph of the architecture. An FPGA is basically represented by the floor-planning of different blocks and the routing graph connecting these blocks. The FPGA VHDL model generator

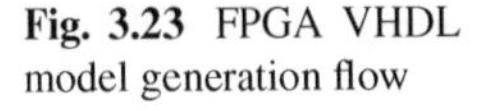

Fig. 3.23 FPGA VHDL model generation flow

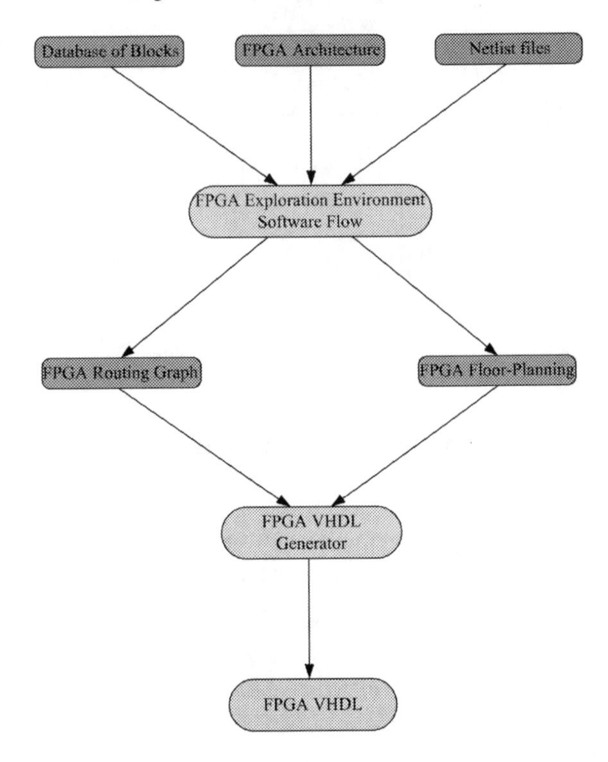

uses FPGA floor-planning and FPGA routing graph to generate the FPGA VHDL model.

3.6.2 FPGA VHDL Model Generation

As shown in Fig. 3.23, the VHDL model of an FPGA is generated by using FPGA floor-planning and FPGA routing graph. An FPGA architecture may contain different kinds of blocks including I/Os, CLBs. The FPGA floor-planning gives the position of different blocks on the FPGA. These blocks are interconnected through a routing network. The routing network of the FPGA is represented by a routing graph. An FPGA routing graph contains nodes that are connected through edges; nodes represent a wire, and an edge represent the connections between different wires. A wire in the routing graph can be an input or output pin of a block, or a routing wire of the routing network.

The VHDL model generation using routing graph is explained with the help of a small example as shown in Fig. 3.24. Figure 3.24a shows a generalized full cross bar switch block which has four inputs and four outputs. The routing graph for this switch block is shown in Fig. 3.24b. This routing graph can be parsed to generate its

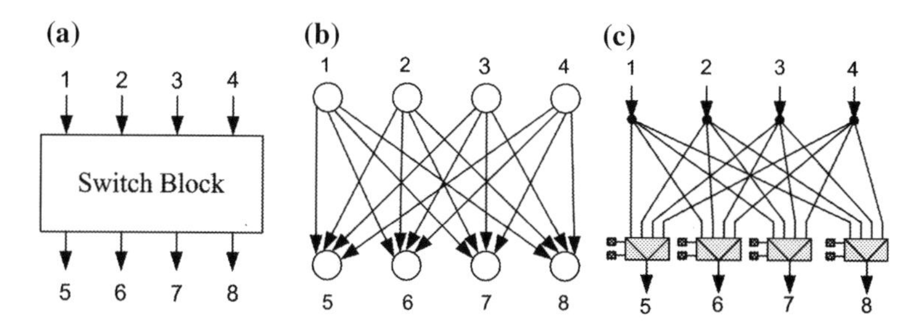

Fig. 3.24 FPGA VHDL model generation from routing graph. **a** Full cross-bar switch block, **b** routing graph for FPGA switch block, **c** physical representation for FPGA switch box

physical representation. The physical representation of FPGA switch block is shown in Fig. 3.24c. If a node is driven by more than one nodes, a multiplexor along with the required SRAMs is used to drive multiple nodes to the receiver node. If a node is driven by only a single node, a buffer is used to drive the receiver node. The physical representation of the routing graph is later translated to a VHDL model. This VHDL model is generated using a symbolic standard cell library, SXLIB [9].

The FPGA routing graph is parsed to generate its physical representation. If a receiver wire is driven by two or more wires, a multiplexor of appropriate size is connected to the receiver wire. If the receiver wire is being driven by only one wire, a buffer is used to connect the driver wire to the receiver wire. When a multiplexor is inserted, the SRAM bits required by the multiplexor are also declared along. The multiplexors and buffers belong to the same tile to which the receiver wire belongs. The SRAM bits connected to the multiplexor also belong to the same tile to which the multiplexor belongs.

The IO and logic block instances are also declared. The logic block can be a soft-block such as a CLB, or a hard-block such as multiplier or adder etc. The input and output pins of these blocks are already represented in the routing graph. Thus, these blocks are automatically linked to physical representation of an FPGA. These blocks are declared in their respective tiles. The VHDL model of these blocks is provided along with the architecture description. The SRAMs used by any logic blocks are also placed in the same way as the SRAMs of routing network are placed.

3.6.3 FPGA Layout Generation

The FPGA VHDL model is generated using a symbolic standard cell library, SXLIB [9]. The generated VHDL model is synthesized to 130 nm standard cell library of STMicroelectronics, and then passed to Cadence encounter for layout generation. The fanout loads are fixed through automatic buffer insertion and gate resizing. The default parameters of encounter enforce at least 5% space reserved for empty space (fillers). FPGA layout generation uses these default parameters. If there is too much congestion in the chip, more fillers can be inserted to facilitate routing of a chip.

3.7 Summary and Conclusion

In this chapter, a brief overview of homogeneous mesh-based and tree-based architectures is given. Separate environments are designed for the exploration of the two architectures. Exploration of the architectures starts with a detailed architecture description mechanism and once the architecture is defined, netlists are placed and routed on the architecture using a specifically designed software flow. The new feature of this software flow is that now it can use both uni-directional or bi-directional routing networks for mesh-based architecture. Experiments are performed to evaluate the two architectures and results show that for a set of 16 MCNC benchmarks, on average, tree-based architecture is 29% better than uni-directional mesh-based architecture. A generalized hardware generation method for mesh-based and tree-based FPGA architectures is also presented. The method is automated as it is directly integrated with the exploration environments of both architectures and it takes into account the characteristics that are possessed by different exploration techniques of both architectures. Further the method presented here is generalized in the sense that it can be applied to either homogeneous or heterogeneous architectures.

Results presented in this chapter are for homogeneous architectures (i.e. architectures containing only CLBs and I/Os). Environments presented in this chapter are valid only for homogeneous architecture. However, they act as a stepping stone for a number of aspects that are designed and explored in the remaining chapters of this work. In the next chapter the two architectures and their respective environments are modified to support a heterogeneous mixture of blocks. Heterogeneity in FPGAs has become increasingly important as it gives them advantages in terms of area, speed and power consumption over their homogeneous counterparts. In the next chapter a number of techniques are explored for both mesh-based and tree-based architectures and their effect is evaluated by placing and routing a number of heterogeneous benchmarks on the two architectures. The work and some of the results presented in this chapter are also published in [132].

Chapter 4
Heterogeneous Architectures Exploration Environments

During past few years, the advancement in process technology has resulted in a great increase in the capacity of FPGAs; the devices which were once small have now become large and are used to implement complete designs. Increase in the capacity of FPGAs has allowed their transition from devices that once contained only homogeneous blocks to the devices that now contain a mixture of blocks ranging from soft blocks (e.g. Configurable Logic Blocks) to hard-blocks like multipliers, adders, RAMs etc. The use of hard-blocks in FPGAs has resulted in an improved overall efficiency and now they are used for large and complex applications.

This chapter presents a new exploration environment for tree-based heterogeneous FPGA architecture. This environment is based on the environment discussed in previous chapter. The environment of the previous chapter is modified so that an architecture description mechanism allows to define various architectural parameters including definition of new heterogeneous blocks, the level where they are located and their arity (i.e. number of blocks per cluster). Once the architecture is defined, a software flow partitions and routes the target netlist on the architecture. The partitioning and routing tools are modified to incorporate a mixture of heterogeneous blocks in the architecture. A mesh-based heterogeneous exploration environment, initially presented in [92], is also explored and enhanced in this chapter. This environment is an extended version of homogeneous exploration environment of mesh-based architecture presented in previous chapter. Different floor-planning techniques are explored for mesh-based architecture using different sets of benchmarks and results of those benchmarks are compared with results of tree-based architecture.

4.1 Introduction and Previous Work

During recent past, embedded hard-blocks (HBs) in FPGAs (i.e. heterogenous FPGAs) have become increasingly popular due to their ability to implement complex applications more efficiently as compared to homogeneous FPGAs. The work

U. Farooq et al., *Tree-Based Heterogeneous FPGA Architectures*,
DOI: 10.1007/978-1-4614-3594-5_4,

in [123] shows that the use of embedded memory in FPGA improves its density and performance. Authors in [19] have incorporated floating point multiply-add units in the FPGA and have reported significant area and speed improvements over homogeneous FPGAs. In [58] a virtual embedded block (VEB) methodology is proposed that predicts the effects of embedded blocks in commercial FPGA devices; and it has shown that the use of embedded blocks causes an improvement in area and speed efficiencies. Also authors in [52] and [118], suggest the use of embedded blocks in FPGAs for better performance regarding complex scientific applications. The work in [72] shows that the use of HBs in FPGAs reduces the gap between ASIC and FPGA in terms of area, speed and power consumption. Some of the commercial FPGA vendors like Xilinx [126] and Altera [13] are also using HBs (e.g. multipliers, memories and DSP blocks) in their architectures.

Almost all the work cited above considers mesh-based FPGAs as the reference architecture where HBs are placed in fixed columns; these columns of HBs are interspersed evenly among columns of configurable logic blocks (CLBs). The main advantage of mesh-based, fixed-column heterogeneous FPGA lies in its simple and compact layout generation. However, the column-based floor-planning of FPGA architectures limits each column to support only one type of HB. Due to this limitation, the architecture is bound to have at least one separate column for each type of HB even if the application or a group of applications that is being mapped on it uses only one block of that particular type. This can eventually result in the loss of precious logic and routing resources. This loss can become even more severe with the increase in number of types of blocks that are required to be supported by the architecture.

Although, significant amount of research has already been done regarding mesh-based heterogeneous FPGA architectures; no work has been done yet in this domain for tree-based heterogeneous FPGA architectures. Contrary to mesh-based architectures where logic and routing resources are arranged in an island style, tree-based architecture is a hierarchical architecture where logic and routing resources are arranged in a multilevel clustered structure. So, in this chapter we present a new exploration environment for tree-based heterogeneous FPGA architectures. Different techniques are explored to optimize the use of logic and routing resources of the architecture. Further, in this chapter, an exploration environment for mesh-based heterogeneous FPGA architecture is described [92]. The environment for mesh-based architecture is jointly developed by authors of this book and a previous student of our research team [94]. Contrary to existing environments of mesh-based architecture that use a fixed floor-planning technique, this environment automatically optimizes the floor-planning of hard-blocks. Also, unlike previous research [72, 123] that mainly compares mesh-based heterogeneous FPGA architectures with their homogeneous counterparts, this chapter presents a detailed comparison between different architectural techniques of heterogeneous mesh-based and tree-based architectures.

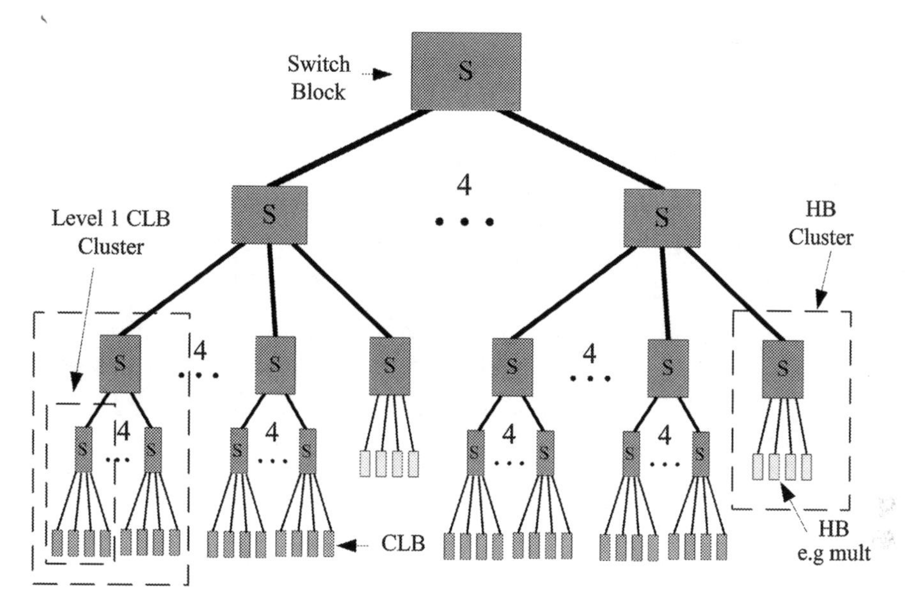

Fig. 4.1 Generalized tree-based heterogeneous FPGA architecture

4.2 Reference Heterogeneous FPGA Architectures

This section gives basic overview of the two heterogeneous FPGA architectures. These FPGA architectures are based on the architectures that are described in Chap. 3 and modifications are made in them so they can support a mixture of heterogenous blocks.

4.2.1 Heterogeneous Tree-Based FPGA Architecture

A tree-based heterogeneous architecture [45] is a hierarchical architecture having unidirectional interconnect. In tree-based heterogeneous architecture CLBs, I/Os and HBs are partitioned into a multilevel clustered structure where each cluster contains sub clusters and switch blocks allow to connect external signals to sub-clusters. Figure 4.1 shows generalized example of a four-level, arity-4, tree-based architecture. The arity of the architecture is basically defined as the number of CLBs in the base-cluster of the architecture and it is respected as we move towards the top of the hierarchy. However, in a heterogeneous architecture we may have a mixed arity. For example in Fig. 4.1, first two levels have arity 4, third level has arity 5 and fourth level has arity 4. But in order to keep the convention, we refer it as arity 4 architecture because its base-cluster contains four CLBs. In a heterogenous tree-based architecture, CLBs are placed at the bottom of hierarchy whereas HBs can be

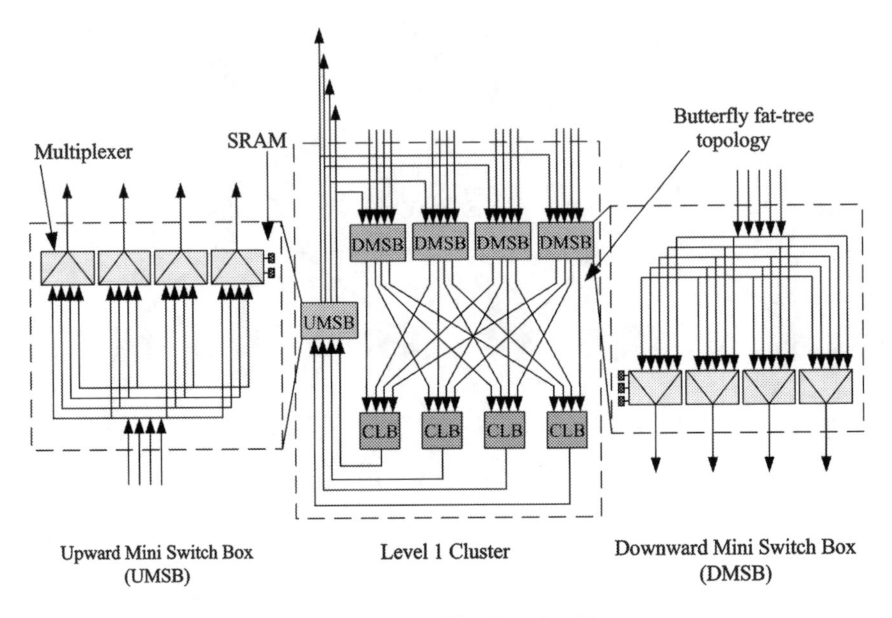

Fig. 4.2 Detailed interconnect of base-cluster of tree-based architecture

placed at any level of hierarchy to meet the best design fit. For example, in Fig. 4.1 HBs are placed at level 2 of hierarchy.

4.2.1.1 The Interconnect Network

Similar to homogeneous architecture, tree-based heterogeneous architecture contains two unidirectional, single length, interconnect networks: a downward network and an upward network. As we move towards the top, signal bandwidth grows in both networks and it is maximum at the top of hierarchy. Downward network is based on butterfly fat tree topology and allows to connect signals coming from other clusters to its sub-clusters through a switch block. The upward network is based on hierarchy and it allows to connect sub-cluster outputs to other sub-clusters in the same cluster and to clusters in other levels of hierarchy. A detailed base-cluster example of two interconnect networks is shown in Fig. 4.2. In this figure, base-cluster contains four CLBs where each CLB contains one LUT with 4 inputs and one output. It can be seen from the figure that switch blocks are further divided into downward and upward mini switch boxes (DMSBs and UMSBs). These DMSBs and UMSBs are unidirectional full cross bar switches that connect signals coming into the cluster to its sub-clusters and signals going out of a cluster to other clusters of hierarchy.

Because of the homogeneity at all the levels, in a tree-based homogeneous architecture, the number of DMSBs in a switch block of a cluster at level ℓ are equal to number of inputs of a cluster at level $\ell - 1$. Similarly, number of UMSBs in a

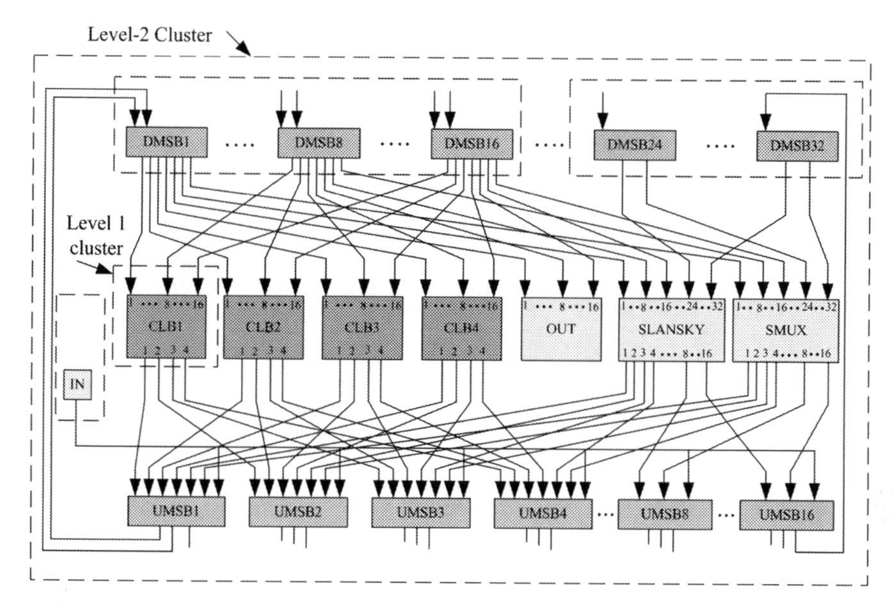

Fig. 4.3 A two level tree-based heterogeneous FPGA architecture

cluster at level ℓ are equal to number of outputs of a cluster at level $\ell - 1$. However, this rule is slightly changed in case of tree-based heterogeneous architectures. In a tree-based heterogeneous architecture, since, there can be clusters at a level with different number of inputs and outputs. So, in this case the number of DMSBs in a cluster at level ℓ are equal to the the highest number of inputs of a cluster at level $\ell - 1$ and number of UMSBs in a cluster at level ℓ are equal to the highest number of outputs of a cluster at level $\ell - 1$. In this way all the inputs of a cluster with maximum inputs can be reached by DMSBs of of upper level cluster and all the outputs of the cluster with maximum outputs can be connected to UMSBs of upper level cluster. By applying this rule we ensure high flexibility and hence improved routability. Also by using this principle we remain inline with the connection patterns of homogeneous architecture explained in previous chapter. This rule can be further explained with the help of example shown in Fig. 4.3. In this figure a two level tree-based heterogeneous architecture is shown. This architecture contains four CLB clusters (detail shown in Fig. 4.2) and two hard-block clusters. We can see that SLANSKY and SMUX (two hard-blocks used by one of the benchmarks) clusters have highest number of inputs among all the clusters of level 1 which is 32. So the number of DMSBs of a cluster at level 2 are equal to 32 and same rule is applied to determine the number of UMSBs at that level. Once the number of DMSBs and UMSBs of a cluster are determined, the inputs per DMSB and outputs per UMSB are determined and connected to the inputs and outputs of lower and upper level clusters in a similar manner as that of tree-based homogeneous architecture.

4.2.1.2 Interconnect Depopulation

Generally, in a tree-based architecture, the interconnect bandwidth grows from bottom to top. However, the number of signals entering into and leaving from the cluster situated at a particular level of a tree-based architecture can be varied depending upon the netlist requirement. The signal bandwidth of clusters is controlled using Rent's rule [74] which is easily adapted to tree-based heterogeneous architecture. This rule states that

$$IO = \left(\underbrace{k.n^{\ell}}_{L.B(p)} + \underbrace{\sum_{x=1}^{z} a_x.b_x.n^{(\ell-\ell_x)}}_{H.B(p)} \right)^{p} \tag{4.1}$$

where

$$H.B(p) = \begin{Bmatrix} 0 & if(\ell - \ell_x < 0) \\ a_x.b_x.n^{(\ell-\ell_x)} & if(\ell - \ell_x \geq 0) \end{Bmatrix} \tag{4.2}$$

In Eq. 4.1 ℓ is a tree level, n is the arity size, k is the number of in/out pins of a LUT, a_x is the number of in/out pins of a HB of type x, ℓ_x is the level where HB is located, b_x is the number of HBs at the level where it is located, z is the number of types of HBs supported by the architecture and IO is the number of in/out pins of a cluster at level ℓ. Since there can be more than one type of HBs, their contribution is accumulated and then added to the $L.B(p)$ part of Eq. 4.1 to calculate p. The value of p is a factor that determines the cluster bandwidth at each level of the tree-based architecture and it is averaged across all the levels to determine the p for the architecture. Normally the value of p is either ≤ 1. However, in particular cases, the initial value of p can be >1 (for details please refer to Sect. 5.6).

4.2.2 Heterogeneous Mesh-Based FPGA Architecture

A mesh-based heterogeneous FPGA is represented as a grid of equally sized slots which is termed as slot-grid. Blocks of different sizes can be mapped on the slot-grid. A block can be either a soft-block like a configurable logic block (CLB) or a hard-block like multiplier, adder, RAM etc. Each block (CLB or a HB) occupies one or more slots depending upon its size. The architecture used in this work is a VPR-style (Versatile Place and Route) [81] architecture that contains CLBs, I/Os and HBs that are arranged on a two dimensional grid. In order to incorporate HBs in a mesh-based FPGA, the size of HBs is quantized with size of the smallest block of the architecture i.e. CLB. The width and height of an HB is therefore a multiple of the width and height of the smallest block in the architecture. An example of such FPGA where CLBs and HBs are mapped on a grid of size 8×8 is shown in Fig. 4.4.

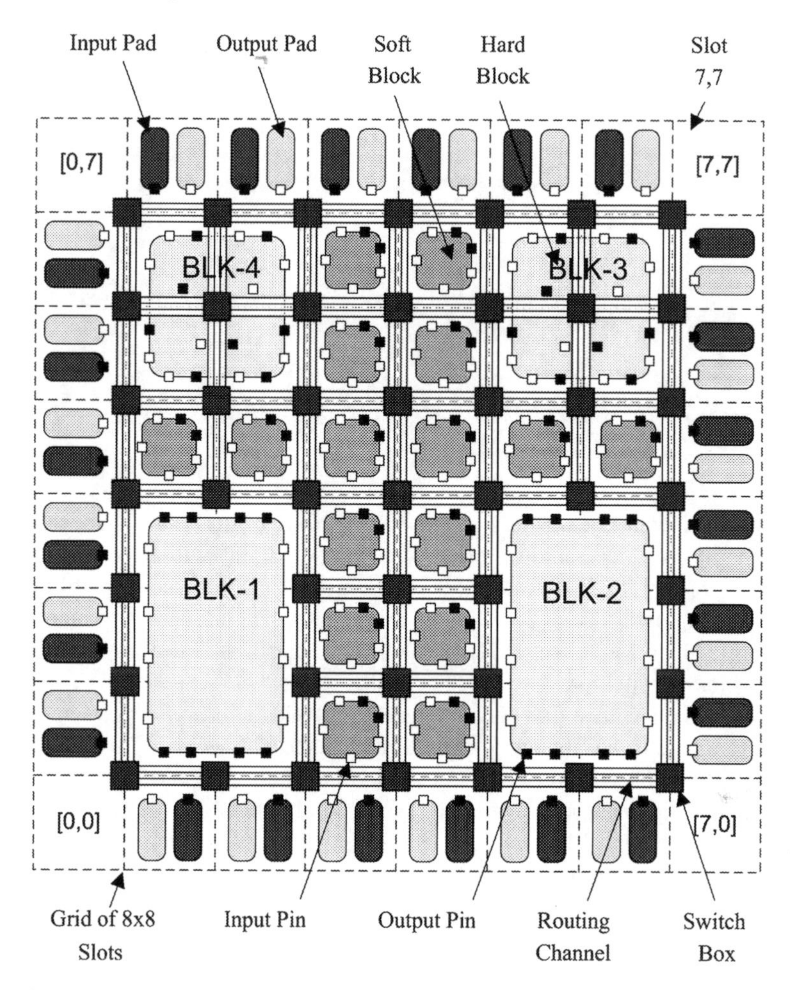

Fig. 4.4 Generalized mesh-based heterogeneous FPGA architecture [94]

In mesh-based FPGA, input and output pads are arranged at the periphery of the slot-grid as shown in Fig. 4.4. The position of different blocks in the architecture depends on the used floor-planning technique. A block (referred as CLB or HB) is surrounded by a uniform length, single driver, unidirectional routing network [77]. The input and output pins of a block connect with the neighboring routing channel. In the case where HBs span multiple tiles, horizontal and vertical routing channels are allowed to cross them [19]. In case of mesh-based heterogeneous FPGA, the detailed interconnect of a CLB with surrounding routing network remains same as homogeneous FPGA (already detailed in Fig. 3.2). For HBs, their input and output pins are connected to the surrounding network in a similar manner as CLBs except that CLBs occupy only one slot and HBs may occupy multiple slots (refer to Fig. 4.4).

In order to explore the two architectures, architecture exploration environments are developed for both of them. Architecture exploration environment of both mesh-based and tree-based architectures starts with respective architecture description. Once the architecture is defined, netlists are placed and routed on both architectures using a software flow. Different steps that are involved in the exploration are detailed in following sections.

4.3 Architecture Description

4.3.1 Architecture Description of Heterogeneous Tree-Based Architecture

Different architecture parameters of the tree-based heterogeneous FPGA architecture are defined using an architecture description file. Some of these parameters are shown in Table 4.1. The parameter Nb_Levels defines the total number of levels of the architecture. Nb_Block_Types parameter defines the total number of types that are supported by the architecture. By default, a tree-based architecture supports two types of blocks which are logic blocks (CLBs or soft blocks) and I/Os. For heterogeneous architectures, however, the types of blocks may vary depending on the netlist requirements that are being implemented on the architecture. The architecture is quite flexible in this sense and it can support any number of block types that can be placed at different levels of hierarchy in order to have a best design fit. In our description mechanism, the architecture description starts with the specification of I/O blocks and once it is done, the rest of the architecture is defined repeatedly using the parameters of lines 5–10 of Table 4.1. These parameters include level number being defined, number of sub-cluster types supported by each cluster, number of sub-clusters contained in each cluster and number of inputs/outputs of the cluster. Definition of clusters starts from bottom level of the architecture and it goes to top until all the levels of the architecture are specified. Once cluster definition is over, binary search parameter is either set to be true or false. If binary search parameter is false, no architecture optimization is performed and the netlist is routed using the given cluster bandwidth. However if this parameter is true, then an architecture optimization is performed using the specified optimization approach which can be bottom_up, top_down or random. Details regarding these optimization approaches are already given in Sect. 3.5.1.

Once cluster definition of all the levels is over, different types of blocks that are supported by the architecture can be defined using Define_Block parameter. Different parameters that are used for the definition of a block are shown in Table 4.2. In a tree-based architecture, definition of a block starts with the name of the block. The parameter "Area" gives the area of the block which is later used for the area calculation of the architecture. Other parameters include the number of input/output pins, the level where the block is located, the arity (i.e. number of blocks per cluster) of the

Table 4.1 Architecture description file parameters of tree-based architecture

	Name	Description
1.	Nb_Levels	Total number of levels in the architecture
2.	Nb_Block_Types	Number of block types that are supported by the architecture
3.	In_Blocks	The level of the cluster and the number of inputs per cluster of the input block
4.	Out_Blocks	The level of the cluster and the number of outputs per cluster of the output block
5.	Level	The level ℓ of the architecture
6.	Nb_Cluster_Type	Number of sub-cluster types supported by a cluster of level ℓ
7.	Arity	Number of sub-clusters of each type supported by a cluster of level ℓ
8.	Nb_Inputs_Per_Cluster	Number of inputs per cluster of each type
9.	Nb_Outputs_Per_Cluster	Number of outputs per cluster type
10.	End_Level	Completes the definition of level ℓ
11.	Optimization	Binary search flag set either true or false
12.	Optimization_approach	Specified as either bottom_up, top_down or random
13.	Define_Block blk	Block definition (See Table 4.2)

Table 4.2 Block definition in tree-based architecture

Definition	Description
Define_Block	
Block_Name	Name of the block
Area	Area of the block
Nb_Inputs	Number of inputs of the block
Nb_Outputs	Number of outputs of the block
Level_Number	Level number where the block is located
Arity	Number of blocks per cluster
Pin_Input	Name and the class number of input pins of the block
Pin_Output	Name and the class number of output pins of the block
End_Define_Block	

block and the definition of its input and output pins. While defining I/O pins of a particular block (logic-block or a hard-block), unique class number are assigned to each block pin to ensure the appropriate routing of the netlist that is mapped on the architecture. An example of the architecture description file that we use to construct the architecture is shown in Fig. 4.5.

Fig. 4.5 An example of architecture description file for tree-based FPGA architecture

```
New_device

Nb_levels:                                              2
Nb_block_types:                                         2

In_iobs      level        1        number               4
Out_iobs     level        1        number               4

End_device

Architecture_description:

.level                                                  0

        Arity:                                          4
        nb_cluster_type:                                1

        new_cluster_type:
                nb_inputs_per_cluster:                  16
                nb_outputs_per_cluster:                 4
        end_cluster_type:

End_level:                                              0

.level                                                  1

        Arity:                                          4
        nb_cluster_type:                                1

        new_cluster_type:
                nb_inputs_per_cluster:                  0
                nb_outputs_per_cluster:                 0
        end_cluster_type:

End_level:                                              1

End_architecture_description

Optimization:                                           true
Optimization_algo:                                      bottom_up
T-driven:                                               false

Define_block

Block_name:                                             CLB
Area:                                                   58500
Nb_inputs:                                              4
Nb_outputs:                                             1
Level_number:                                           0
Arity:                                                  4
Pin_input:                                      i0      0
Pin_input:                                      i1      0
Pin_input:                                      i2      0
Pin_input:                                      i3      0
Pin_output:                                     q0      1

End_define_block
```

Table 4.3 Architecture description file parameters of mesh-based architecture

Name	Description
1. Nx num	Slots in the slot-grid in X direction (num >1)
2. Ny num	Slots in the slot-grid in Y direction (num >1)
3. Input_Rate	Number of input pads in each slot on the periphery of slot-grid
4. Output_Rate	Number of output pads in each slot on the periphery of slot-grid
5. Channel_Type T	T is unidirectional or bidirectional
6. Binary_Search F	Binary search flag (F is true or false)
7. Channel_Width num	Channel width if Binary_Search = false (num > 1)
8. Channel_Width_Min num	Minimum channel width if Binary_Search = true (num > 1)
9. Channel_Width_Max num	Maximum channel width if Binary_Search = true (num > 1)
10. Set_Block blk X Y	Place a block named blk at a slot position (X,Y) of slot-grid
11. Set_Block_Auto blk N	Place N instances of blk on first available position of slot-grid
12. Fix_Block_Positions F	The Blocks are movable or fixed (F is true or false)
13. Block_Jump F	If Blocks are moveable, blocks can be moved (F is true or false)
14. Block_Rotate F	If Blocks are moveable, blocks can be rotated (F is true or false)
15. Column_Move W s	If Blocks are moveable, a column can be moved (W is width of the column, s is the starting horizontal slot position of column)
16. Define_Block blk	Block definition (See Table 4.4)

4.3.2 *Architecture Description of Heterogeneous Mesh-Based Architecture*

Architecture description file of mesh-based FPGA architecture comprises of a number of parameters that are used to construct the architecture. Few major architecture description parameters are shown in Table 4.3. The parameters Nx and Ny define the size of the slot-grid. Channel_Type is used to select a unidirectional mesh [77] or a bidirectional mesh [120] routing network. The channel width of the routing network is either set to a constant value (using the parameter Channel_Width), or a binary search algorithm searches a minimum possible channel width between minimum (Channel_Width_Min) and maximum (Channel_Width_Max) channel width limits. In case of unidirectional mesh, the channel width remains in multiples of 2. The position of blocks can be set to an absolute position on the slot-grid (by using the parameter Set_Block). This parameter takes the name of the block and the position on the slot-grid where it should be placed. Another option to place blocks on the slot-grid is by using the parameter Set_Block_Auto. This parameter automatically places N copies of a block on the first available position on the slot-grid. The blocks on the slot-grid can be either fixed to an initial position or set as moveable (by using the parameter Fix_Block_Positions). In case the blocks are moveable, the placer can refine their position on the slot-grid. The parameter Block_Jump allows the placer to move blocks on the slot-grid. The parameter Block_Rotate allows the placer to rotate blocks at their own axis. The parameter Column_Move allows to move a complete column from one position to another. Column_Move parameter requires the width

Table 4.4 Block definition in mesh-based architecture

Definition	Description
Define_Block	
X_Slots	Number of slots occupied by the block in horizontal direction
Y_Slots	Number of slots occupied by the block in vertical direction
Rotate	A flag set true or false to allow or restrict the rotation of block
Area	Area of the block
Pin_Input	Name, position, class and the direction of input pin of block
Pin_Output	Name, position, class and the direction of output pin of block
End_Define_Block	

of column, W (i.e number of slots as column width), and the starting horizontal slot position of the column. All the blocks in a column must be within the boundaries of the column. This parameter can be repeated if multiple columns are required to be moved.

A new block can be defined in the architecture description file using the Define_Block parameter. The block definition parameters are shown in Table 4.4. Each block is given a name, a size (number of slots occupied), a rotation flag and input/output pins. The rotation flag allows the rotation of individual block by the placer (significance of rotation of a block is explained in Sect. 4.4.3). This rotation flag permits to turn off the rotation of a particular type of block when the global rotation is turned on. Each pin of the block is given a name, a class number, a direction and a slot position on the block to which this pin is connected. An example of the architecture description file that is used to construct a mesh-based FPGA architecture is shown in Fig. 4.6.

4.4 Software Flow

Once the FPGA architectures are defined using their respective architecture description mechanisms, different netlists (benchmarks) are placed and routed on them using a software flow. The software flow used for the exploration of two architectures is shown in Fig. 4.7. The software flow is mainly divided into two parts: first part deals with synthesis and conversion of netlist to .net format while remaining flow deals with the architecture exploration. It can be seen from the figure that netlist synthesis involves a number of steps before it can be placed and routed on the FPGA architecture. These steps are common for both mesh-based and tree-based architectures and they convert the netlist from .vst format to .net format. The netlist in .vst format is obtained using VASY [62] that converts VHDL file to structured VHDL (.vst). Normally a netlist in VST format is composed of traditional standard cell library instances and hard-block instances. The VST2BLIF tool converts the VST file to BLIF format. Later, PARSER-1 removes all the instances of hard-blocks and passes the remaining netlist to SIS [102] for synthesis into 4 input Look-Up Table format.

```
Nx:                                 4
Ny:                                 4

In_rate:                            1
Out_rate:                           1

Interconnect:                       uni_mesh
Binary_search:                      true
Network_width:                      6
Binary_search_min:                  0
Binary_search_max:                  40

Set_automatic           CLB         16

Block_jump:                         true
Block_rotate:                       true
Column_move:                        false

Define_block

Block_name:                         CLB
X_slots                             1
Y_slots                             1
Rotate                              true
Area:                               58500
Pin_input:          i0    0    0 0  PIN_LEFT
Pin_input:          i1    0    0 0  PIN_RIGHT
Pin_input:          i2    0    0 0  PIN_TOP
Pin_input:          i3    0    0 0  PIN_BOTTM
Pin_output:         q0    1    0 0  PIN_TOP PIN_RIGHT

End_define_block
```

Fig. 4.6 An example of architecture description file for mesh-based FPGA architecture

All the dependence between hard-blocks and remaining netlist is preserved by adding new input and output pins to the main netlist. SIS generates a network of LUTs and Flip-Flops, which are later packed into CLBs through T-VPACK [120]. T-VPACK generates a netlist in NET format and then PARSER-2 adds all the removed hard-blocks into this netlist. It also removes all the inputs and outputs temporarily added by PARSER-1. This final netlist in NET format, containing CLBs and hard-blocks, is then placed and routed separately on mesh-based and tree-based architectures. In this flow SIS is used for synthesis which we want to replace with ABC [21] in future. Few of the major components of the software flow are detailed below.

4.4.1 Parsers

The output generated by VST2BLIF tool is a BLIF file containing input and output port instances, gates belonging to a standard cell library, and hard-block instances (which are represented as sub circuits in BLIF format). This BLIF file is passed to

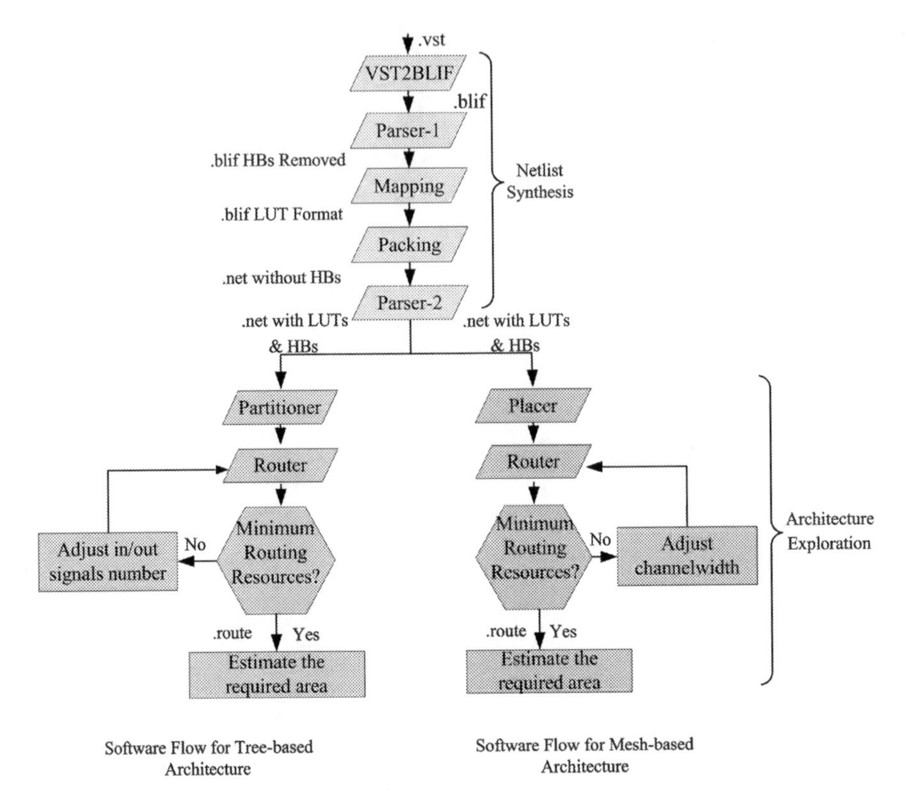

Fig. 4.7 Software flow

SIS [102] for synthesis into LUT format. However, the hard-blocks in the BLIF file are not required to be synthesized. So, the main aim of PARSER-1 is to remove hard-blocks from BLIF file in such a way that all the dependence between the hard-blocks and the remaining netlist is preserved. After synthesis and packing, PARSER-2 will add all the removed hard-blocks in the netlist.

Figure 4.8 shows five different modifications performed by PARSER-1 before removing hard-block instances from the BLIF file. These cases are described as below:

1. Figure 4.8a shows a hard-block whose output pin is connected to the input pin of gate. The output pin of hard-block is detached from the input pin of gate. The detached signal is added as the input pin of main circuit, as shown in Fig. 4.8b.
2. All the output pins of main circuit that are connected by the output pins of hard-block (as shown in Fig. 4.8c) are connected to zero gates (as shown in Fig. 4.8d). This is because, when hard-block is removed, these main circuit outputs do not remain stranded.

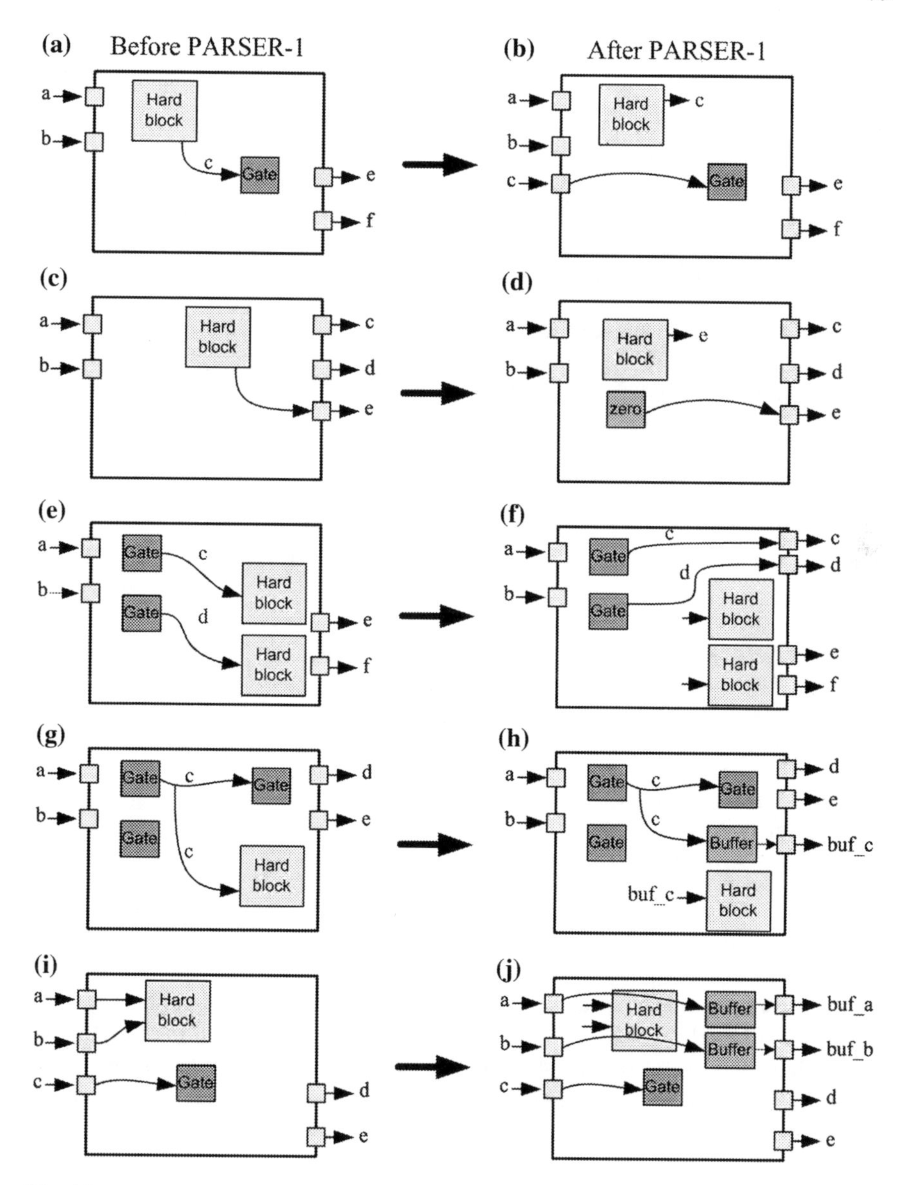

Fig. 4.8 Netlist modifications done by PARSER-1 [94]

3. All the output pins of gates connected only to the input pins of hard-blocks (shown in Fig. 4.8e) are added as the output pins of main circuit (as shown in Fig. 4.8f).
4. For all the output pins of gates connected to the input pins of hard-blocks and also to the input pins of gates (as shown in Fig. 4.8g), add a buffer to this gate

output. The buffered output is added as the output of main circuit. The name of the buffered output should be replaced in all the input pins of hard-blocks (as shown in Fig. 4.8h).

5. Figure 4.8i shows the input pins of main circuit that are connected only to input pins of hard-blocks. After the removal of hard-blocks these inputs will remain stranded and will eventually be removed by SIS. To avoid their removal, these input pins are retained by adding buffers to them, and adding the buffered outputs to the main circuit outputs.

After performing the above changes, PARSER-1 removes all the hard-blocks from the BLIF file. The BLIF file without hard-blocks is passed to SIS, which converts them to LUTs and Flip-Flops. T-VPACK packs LUTs and Flip-Flops together into CLBs. Next, PARSER-2 performs the following changes

1. Adds hard-blocks in the netlist file which is generated by T-VPACK.
2. Removes the "Main circuit input" and "Main circuit outputs" added by PARSER-1.
3. Removes all the zero gates (represented as a CLB after SIS) added by PARSER-1.

The final output file contains I/O instances, CLB instances and hard-block instances. This file is then separately placed and routed on mesh-based and tree-based architectures.

4.4.2 Software Flow for Heterogeneous Tree-Based Architecture

4.4.2.1 Partitioning

Once the netlist is obtained in .net format, it is partitioned using a partitioner. The partitioner is based on the one described in previous chapter. Partitioner partitions CLBs, HBs and I/Os into different clusters in such a way that the inter-cluster communication is minimized. By minimizing inter-cluster communication we obtain a depopulated global interconnect network and hence reduced area. Partitioner is based on hMetis [50] platform. hMetis combines Fidducia-Mattheyses (FM) [47] algorithm with its multi phase refinement approach to optimize the partitioning of the netlist. The main objective of partitioner is to reduce communication between different partitions and FM algorithm achieves this objective using a hill-climbing, non greedy, iterative improvement approach. During each iteration, a block with highest gain is moved from one partition to another and then it is locked and it is not allowed to move during remaining time of iteration. After the block is moved, the gain of all of its associated blocks is recomputed and this process continues until all the blocks are locked. At the end of an iteration, total cost is compared to that of previous iteration and the algorithm is terminated when it fails to improve during an iteration. After the netlist is partitioned, it is placed and routed on the architecture.

4.4.2.2 Routing

Once partitioning is done, placement file is generated that contains positions of different blocks on the architecture. This placement file along with netlist file is then passed to another software module called router which is responsible for routing of the netlist. In order to route all nets of the netlist, routing resources of the interconnect structure are first assigned to the respective blocks of the netlist that are placed on the architecture. These routing resources are modeled as directed graph abstraction $G(V, E)$. In this graph the set of vertices V represents the in/out pins of different blocks and the routing wires in the interconnect structure and an edge E between two vertices, represents a potential connection between the two vertices. Router is based on PathFinder [80] routing algorithm that uses an iterative, negotiation-based approach to successfully route all nets in a netlist. In order to optimize the FPGA architecture, a binary search algorithm is used. This algorithm determines the minimum number of signals required to route a netlist on FPGA.

4.4.3 Software Flow for Heterogeneous Mesh-Based Architecture

4.4.3.1 Placement

For mesh-based architecture, the netlist obtained in .net format is placed on the architecture using the placement algorithm that determines the position of different block instances of a netlist on their respective block types on FPGA architecture. The main goal is to place connected instances near each other so that minimum routing resources are required to route their connections. The placer uses simulated annealing algorithm [37, 105] to achieve a placement having minimum sum of half-perimeters of the bounding boxes of all the nets. This placer also optimizes floor-planning of different blocks on the FPGA architecture. Different operation that are performed by the placer are detailed as below.

4.4.3.2 Placer Operations

The placer either moves an instance from one block to another, moves a block from one slot position to another, rotates a block at its own axis, or moves an entire column of blocks. After each operation, the bounding box cost (also called as placement cost) is recomputed for all the disturbed signals. Depending on the cost value and the annealing temperature, the simulated annealing algorithm accepts or rejects the current operation.

The placer performs its operation on "source" and "destination" and the slots occupied by source and destination are termed as source window and destination window respectively. Normally, source window contains one block whereas destination window can contain multiple blocks. An example of source and destination windows is shown in Fig. 4.9a and b respectively. Once the source and destination windows are selected, the move operation is performed if:

1. Destination window does not contain any block that exceeds the boundary of destination window. An example violating this condition is shown in Fig. 4.9c.
2. The destination window does not exceed the boundaries of slot-grid (refer to Fig. 4.9d).
3. Destination window does not overlap source window diagonally (refer to Fig. 4.9g). However if the destination window overlaps source window vertically or horizontally, then horizontal or vertical translation operation is performed. Figure 4.9e shows an example where destination window overlaps source window vertically and Fig. 4.9f shows that the move operation is performed using vertical translation.

However, if above three conditions are not met, the procedure continues until a valid destination window is found. After the selection of source and destination, placer either moves an instance, moves a block, rotates a block, or moves an entire column of blocks. The rotation of blocks is important when the class number assigned to the input pins of a block are different; bounding box varies depending upon the pin positions and their directions. A block can have an orientation of 0°, 90°, 180° or 270°. Figure 4.9h depicts a 90° clock-wise rotation. Multiples of 90° rotation are allowed for all the blocks having a square shape, whereas at the moment only multiples of 180° rotation are allowed for rectangular (non-square) blocks. A 90° rotation for non-square blocks involves both rotation and move operations, which is left for future work.

4.4.3.3 Routing

After the placement of netlist on the FPGA architecture, the exploration environment constructs routing graph for the architecture. Few of the architecture description parameters required for the construction of routing graph are taken from the architecture description parameters. These parameters mainly include the type of routing network (unidirectional or bidirectional), channel width, I/O rate, block types and their pin positions on the block. Other parameters depend on the floor-planning details. These parameters include the position of blocks on the slot-grid and their orientation (0°, 90°, 180° or 270°). After the construction of routing graph, the PathFinder routing algorithm [80] is used to route netlists on the routing architecture. In case a binary search operation is used, routing graph is constructed for varying channel widths; routing is tried for each channel width until a minimum channel width is found.

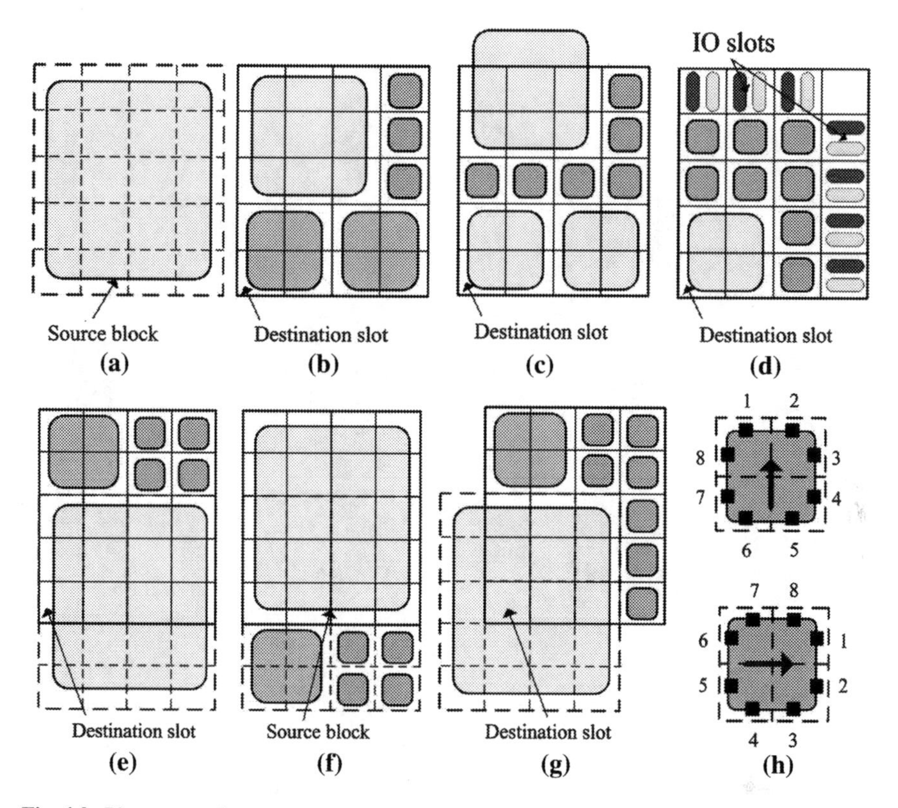

Fig. 4.9 Placer operations

4.4.4 Area Model

Once the optimization of the architecture is over, a generic area model is used to calculate the area of the FPGA (separately for mesh-based and tree-based architecture). The area model is based on the reference FPGA architectures shown in Figs. 4.1 and 4.4 respectively. Area of SRAMs, multiplexors, buffers and Flip-Flops is taken from a symbolic standard cell library (SXLIB [9]) which works on unit Lambda(λ). The area of FPGA is reported as the sum of the areas taken by the switch box, connection box, buffers, soft logic blocks, and hard-blocks.

4.5 Exploration Techniques

Various techniques are explored for both mesh-based and tree-based architectures using the software flow described in Sect. 4.4. A brief overview of different techniques is presented here.

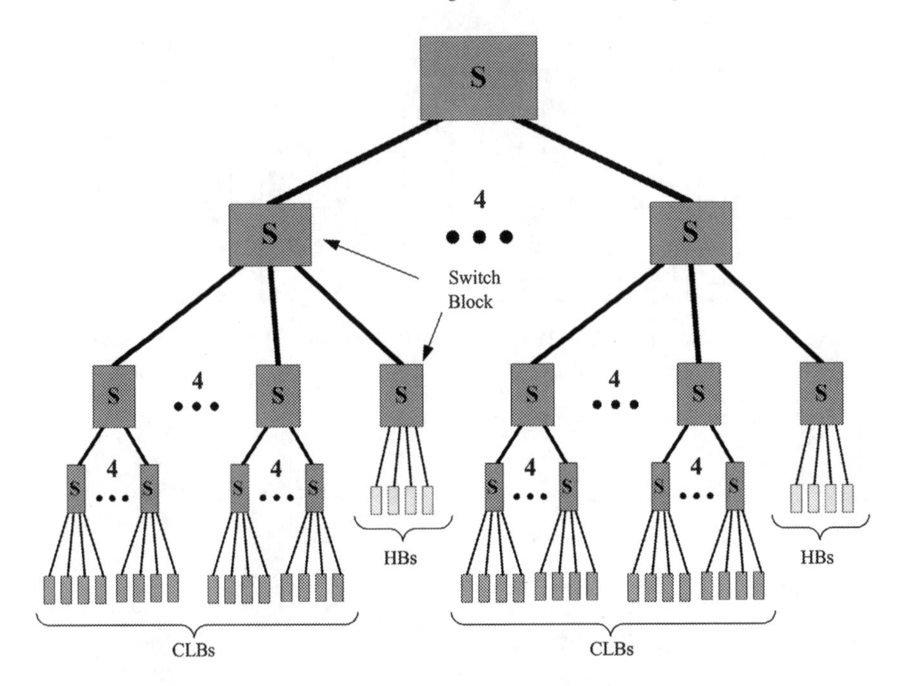

Fig. 4.10 Symmetric tree-based FPGA architecture

4.5.1 Exploration Techniques for Heterogeneous Tree-Based Architecture

Different manual parameters are used in architecture description file to explore two techniques for tree-based architecture. Generalized examples of both techniques are shown in Figs. 4.10 and 4.11 respectively. These techniques are detailed below:

4.5.1.1 Symmetric

A generalized example of first technique is shown in Fig. 4.10. This technique is referred as symmetric (SYM). In this technique clusters of HBs are mixed with those of LBs and HBs can be placed at any level of hierarchy in order to have best design fit. In this technique the symmetry of hierarchy is respected which can eventually result in wastage of HBs and their associated routing resources. For example in Fig. 4.10, it can be seen that this architecture supports four clusters of HBs of a certain type where each cluster contains four HBs. This is because of the fact that this is an arity 4 architecture. However the respect for symmetry of hierarchy may lead to under utilization of HBs and their associated routing resources in the case where a netlist requires less HBs than supported by the architecture.

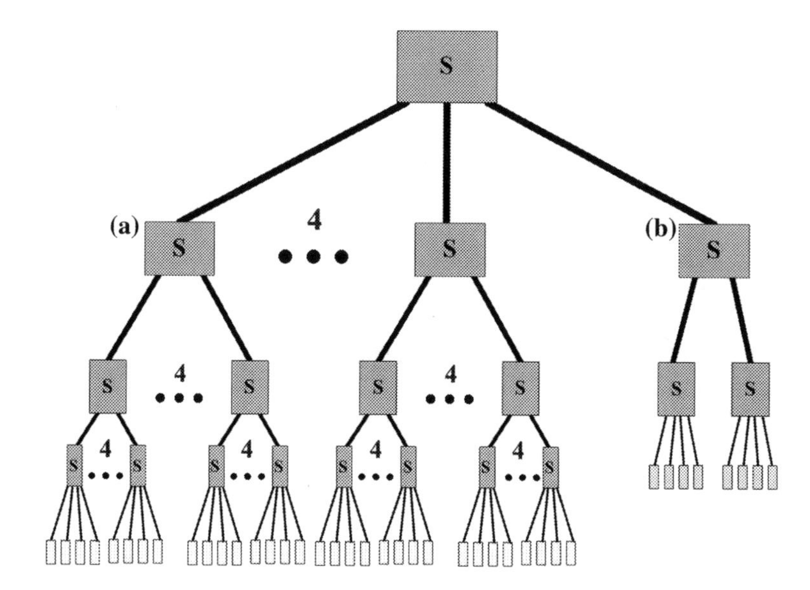

Fig. 4.11 Asymmetric tree-based FPGA architecture

4.5.1.2 Asymmetric

Contrary to the first technique, where the architecture contains only one structure, second technique contains two sub-structures: one sub-structure contains only CLBs while the other contains only HBs. An example of second technique is shown in Fig. 4.11. The main motivation behind this technique is the easy management of logic and routing resources. In this technique both sub-structures are constructed separately and the communication between them is ensured using the switch blocks of their parent cluster. Since the two sub-structures are constructed independently, they do not have to respect the arity of each other; hence leading to more optimized logic and routing resources. This technique is referred as asymmetric (ASYM).

Although this aspect is not fully explored in this work but this type of technique can also be used to exploit arithmetic intensive applications. Arithmetic intensive applications contain a large portion of data-path circuits that contain hard-blocks (e.g. multipliers, adders, memories etc.) that are connected together by regularly structured signals called buses. Conventional FPGAs do not use the regularity of data-path circuits. The regularity of data-path circuits can be exploited by implementing coarse-grain (or bus-based) routing in the sub-structure containing only HBs; routing of the sub-structure containing only CLBs remains unchanged. Exploitation of the regularity of data-path circuits is possible in this technique as the two sub-structures are independent of each other and they communicate with each other only through a parent cluster. By implementing the coarse-grain routing in the sub-structure containing only HBs, the number of SRAMs and switches can be reduced which can eventually lead to smaller area of the architecture.

4.5.2 *Exploration Techniques for Heterogeneous Mesh-Based Architecture*

By using different placer operations, six floor-planning technique are explored for mesh-based architecture. The detail of these floor-planning techniques is as follows:

4.5.2.1 Apart

In this technique, hard-blocks are placed in fixed columns, apart from the CLBs. This technique is shown in Fig. 4.12a and is termed as Apart (A). Such kind of technique can be beneficial for data-path circuits as described by [29]. It can be seen from the figure that if all HBs of a type are placed and still there is space available in the column then in order to avoid wastage of resources, CLBs are placed in the remaining place of column.

4.5.2.2 Column-Partial

Figure 4.12b shows the Column-Partial (CP) technique where columns of HBs are evenly distributed among columns of CLBs.

4.5.2.3 Column-Full

Figure 4.12c shows Column-Full (CF) technique where columns of HBs are evenly distributed among CLBs. Contrary to first and second techniques, whole column contains only one type of blocks. This technique is normally used in commercial architectures and topologically this technique is equivalent to Symmetric and Asymmetric techniques of tree-based FPGA architecture.

4.5.2.4 Column-Move

In this technique, HBs are placed in columns but unlike first three techniques, columns are not fixed, rather they are allowed to move using the column-move operation of placer. This technique is shown in Fig. 4.12d and it is termed as Column-Move (CM).

4.5.2.5 Block-Move

In this technique HBs are not restricted in columns; and they are allowed to move through block move operation. This technique is termed as Block-Move (BM) and it is shown in Fig. 4.12e.

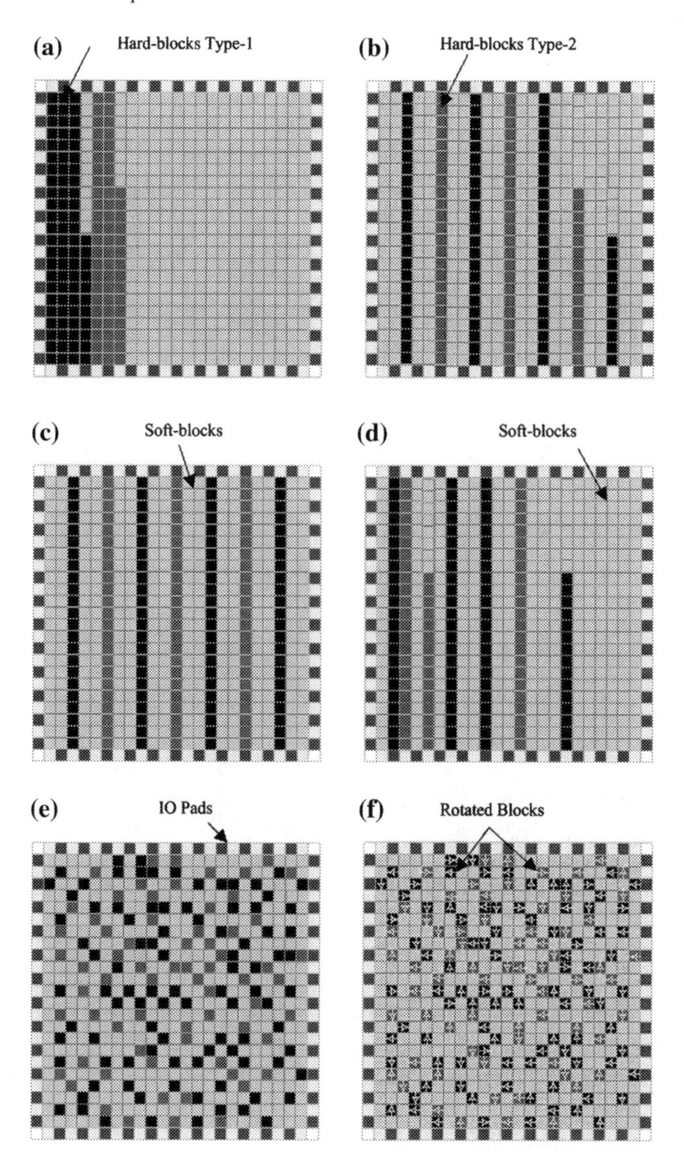

Fig. 4.12 Exploration techniques for mesh-based architecture

4.5.2.6 Block-Move-Rotate

The blocks in this technique are allowed to move and rotate through block move and rotate operations. This floor-planning technique is shown in Fig. 4.12f and it is termed as Block-Move-Rotate (BMR). Among these techniques CF is the technique that is usually used in mesh-based architectures while rest of them are new. Different floor-planning techniques for mesh-based FPGA are explored here because floor-planning can have a major implication on the area of an FPGA. If a tile-based layout is required for an FPGA, similar blocks can be placed in same column. In this way, width of the entire column can be adjusted according to the layout requirements of the blocks placed in a column. On the other hand, if blocks of different types are placed in a column, the width of the column cannot be fully optimized. This is because the column width can only be reduced to maximum width of any tile in that particular column. Thus, some unused space in smaller tiles will go wasted. Such a problem does not arise if a tile-based layout is not required. In such a case, an FPGA hardware netlist can be laid out using any ASIC design flow.

4.6 Experimentation and Analysis

Different techniques of two architectures are explored using the software flow described in Sect. 4.4. Three sets of benchmarks are used for this exploration. The detail about three sets and their selection criteria is described below.

4.6.1 Benchmark Selection

Generally in academia and industry, the quality of an FPGA architecture is measured by mapping a certain set of benchmarks on it. Thus the selection of benchmarks plays a very important role in the exploration of heterogeneous FPGAs. This work puts special emphasis on the selection of benchmark circuits, as different circuits can give different results for different architecture techniques. This work categorizes the benchmark circuits by the trend of communication between different blocks of the benchmark. So, three sets of benchmarks are assembled having distinct trend of inter-block communication. These benchmarks are shown in Tables 4.5, 4.6 and 4.7. Benchmarks shown in Table 4.5 are developed at [79], the benchmarks shown in Table 4.6 are obtained from [36] and the benchmarks shown in Table 4.7 are obtained from [119]. The communication between different blocks of a benchmark can be mainly divided into the following four categories:

1. CLB-CLB: CLBs communicate with CLBs.
2. CLB-HB: CLBs communicate with HBs and vice versa.
3. HB-HB: HBs communicate with HBs.
4. IO-LB/HB: I/O blocks communicate with CLBs and HBs.

In SET I benchmarks, the major percentage of total communication is between HBs (i.e. HB-HB) and only a small part of total communication is covered by the communication CLB-CLB or CLB-HB. On average, in SET I, the HB-HB communication takes up to 80% of the total communication between different instances of the benchmarks (netlists). Similarly, in SET II the major percentage of total communication

Table 4.5 DSP benchmarks SET I

Circuit name	Inputs	Outputs	LUTs (LUT-4)	Mult (8 × 8)	Slansky (16 + 16)	Sff (8)	Sub (8 − 8)	Smux (32:16)	Function
ADAC	18	16	47	–	–	2	–	1	–
DCU	35	16	34	1	1	4	2	2	Discrete cosine transform
FIR	9	16	32	4	3	4	–	–	Finite impulse response
FFT	48	64	94	4	3	–	6	–	Fast fourier transform

Table 4.6 Open core benchmarks SET II

Circuit name	Number of inputs	Number of outputs	Number of LUTs (LUT-4)	Number of multipliers (16 × 16)	Number of adders (20+20)	Function
cf_fir_3_8_8_open	42	18	159	4	3	Finite impulse response (8 bit)
cf_fir_7_16_16	146	35	638	8	14	Finite impulse response (16 bit)
cfft16 × 8	20	40	1,511	–	26	–
cordic_p2r	18	32	803	–	43	Polar to rectangular
cordic_r2p	34	40	1,328	–	52	Rectangular to polar
fm	9	12	1,308	1	19	–
fm_receiver	10	12	910	1	20	–
lms	18	16	940	10	11	–
reed_solomon	138	128	537	16	16	–

is HB-CLB and in SET III, major percentage of total communication is covered by CLB-CLB. Normally the percentage of IO-CLB/HB is a very small part of the total communication for all the three sets of benchmarks.

Table 4.7 Open core benchmarks SET III

Circuit name	Number of inputs	Number of outputs	Number of LUTs (LUT-4)	Number of multipliers (18 × 18)	Function
cf_fir_3_8_8_ut	42	22	214	4	Finite impulse response (8 bit)
diffeq_f_systemC	66	99	1532	4	–
diffeq_paj_convert	12	101	738	5	–
fir_scu	10	27	1366	17	–
iir1	33	30	632	5	Infinite impulse response (16 bit)
iir	28	15	392	5	Infinite impulse response (8 bit)
rs_decoder_1	13	20	1553	13	Decoder
rs_decoder_2	21	20	2960	9	Decoder

4.6.2 Experimental Methodology

Once the benchmarks are selected, they are placed and routed on the two architectures using two different experimental methodologies. An overview of the two methodologies is as follows:

4.6.2.1 Individual Experimentation

In first methodology, experiments are performed individually for each netlist (both for mesh-based and tree-based architectures). The architecture definition, floor-planning, placement/ partitioning, routing and optimization is performed individually for each netlist. Although, such an approach is not applicable to real FPGAs, as their architecture, floor-planning and routing resources are already defined and fixed. However this methodology is useful in order to have detailed analysis of a particular floor-planning technique and usually it is employed to evaluate different parameters of the architecture under consideration. If a generalized architecture is defined for a group of netlists, the netlists with highest logic and routing requirements decide logic and routing resources of the architecture and the behavior of remaining netlists of the group is overshadowed by larger netlists of the group. So, to have more profound analysis, the architecture and floor-planning is optimized for each netlist and the area

Table 4.8 Area of different blocks of three sets

Block name	Inputs	Outputs	Block size (λ^2)
clb	4	1	58,500
mult (8 × 8)	16	16	1,075,250
slansky_16	32	16	306,750
sff_8	8	8	36,000
sub_8	17	8	154,500
smux_16	33	16	36,000
mult (16 × 16)	32	32	1,974,000
adder (20 + 20)	41	21	207,000
mult (18 × 18)	36	36	2,498,300
sram	–	–	1,500
buffer	1	1	1,000
flip-flop	1	1	4,500
mux 2:1	2	1	1,750

of the architecture is calculated individually using the model described in Sect. 4.4.4 and the values shown in Table 4.8. Later average of results of all netlists gives more thorough results.

4.6.2.2 Generalized Experimentation

However, in order to further validate the results, we have also performed experimentation based on the generalized architecture. In this methodology, for mesh-based architecture a generalized minimum architecture is defined for each SET of netlists and the floor-planning is then optimized for this architecture. Generalized floor-planning is achieved by allowing the mapping of multiple netlists on the same architecture where each block of the architecture allows mapping of multiple instances on it, but multiple instances of the same netlist are not allowed. Similarly for tree-based architecture, multiple netlists are partitioned using generalized architecture description where mapping of multiple instances on each block position is allowed but multiple instances of same netlist are not allowed on a single block position. Once generalized floor-planning/partitioning is over, individual netlists are placed and routed separately on both architectures with minimum routing resources that can route any of the netlists of the SET. Since, in this case a generalized architecture is used, optimization of the architecture is not performed for individual netlists.

4.6.3 Results Using Individual Experimentation Approach

Experiments are performed for three sets of benchmarks using the software flow described earlier where 21 benchmarks of three sets are placed and routed individually on the two FPGA architectures using different exploration techniques. Similar to homogeneous architectures experimentation, for tree-based architecture the LUT size is set to be 4 while arity size is set to be 4 too; for mesh-based architecture LUT size is also set to be 4 and CLB size is set to be 1 for both architectures. Since placement cost and channel width of a mesh-based architecture are directly related to its area, we first present the effect of different floor-planning techniques of mesh-based architecture on these two values.

4.6.3.1 Placement Cost and Channel Width Comparison Results

Placement cost results for six floor-planning techniques of mesh-based architecture are shown in Fig. 4.13. In this figure, for each benchmark circuit, placement cost of five floor-planning techniques (Apart (A), Column-Partial (CP), Column-Full (CF), Column-Move (CM) and Block-Move (BM)) of mesh-based FPGA is normalized against the placement cost of Block-Move-Rotate (BMR) technique. Placement cost is the sum of half perimeters of bounding boxes (BBX cost) of all the NETS in a netlist. The results for benchmarks 1–4, 5–13 and 14–21 correspond to SET I, SET II and SET III respectively. For example in Fig. 4.13 column 1 gives normalized results for "ADAC", column 5 gives results for "cf_fir_3_8_8_open" and column 21 gives results for "rs_decoder_2". The avg1, avg2 and avg3 corresponds to the geometric average of these results for SET I, SET II and SET III respectively while avg corresponds to the average of all netlists. As it can be seen from the figure that in general Apart (A) gives the worst and BMR gives the best placement cost results whereas the results of remaining techniques are in between these two techniques. In Apart the average placement cost is higher than the other floor-planning techniques because in this technique columns of hard-blocks are fixed and they are separated from CLBs. Although this kind of floor-planning technique can give good results for data-path circuits, it gives poor placement solution for control path circuits as the columns of HBs are fixed and they are not mixed with CLBs. This situation further aggravates if there are more than one types of HBs that are required to be supported by the architecture. Although columns of HBs are fixed in CF and CP, they give better placement cost results when compared to Apart as in those techniques the columns of hard-blocks are not placed apart rather they are interspersed evenly among CLBs; hence leading to smaller placement costs. BMR gives the best placement cost results because it is the most flexible technique among the six floor-planning techniques. Although the only difference between BM and BMR is that of hard-block rotation, it gives a slight edge to BMR which might lead to smaller bounding box and eventually lower placement costs of the architecture. Figure 4.14 gives the channel width results of the six floor-planning techniques of mesh-based architecture. In this figure, for 21 benchmarks, channel widths of 5 floor-planning techniques are normalized against the channel width of BMR. Similar to the results in Fig. 4.13, BMR gives the best results and Apart gives the worst results. The two figures (i.e. Figs. 4.13 and 4.14) look similar to each other as

1. Both figures give normalized results.

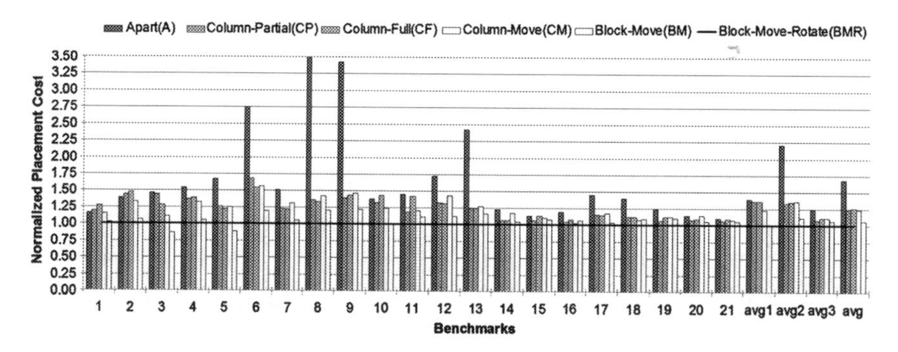

Fig. 4.13 Placement cost comparison between different techniques of mesh-based architecture

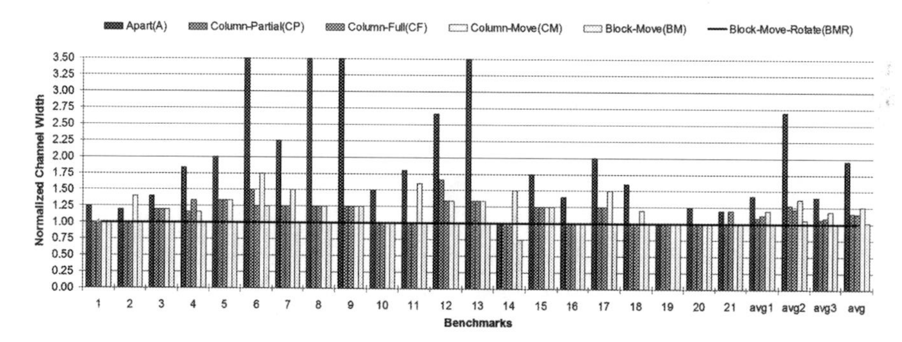

Fig. 4.14 Channel width comparison between different techniques of mesh-based architecture

2. Placement cost and channel width are closely related to each other.

Generally an architecture with higher placement cost indicates a poor placement solution as in this solution instances connected to each other are placed far from each other. A poor placement solution normally leads to higher channel width of the architecture as the instances placed far from each other require more routing resources than the ones placed close to each other. Analysis of the results in Figs. 4.13 and 4.14 show that, on average, CF (the technique used in most of the work cited at the start of the chapter) gives 35, 35 and 11% more placement cost than BMR, for SET I, SET II and SET III benchmark circuits respectively. Figure 4.14 shows that, on average, CF requires 13, 22 and 9% more channel width than BMR for SET I, SET II and SET III respectively.

The reason that BMR gives better placement cost and channel width results when compared to other techniques lies in its flexibility. In four out of five remaining techniques (i.e. Apart, Column-Partial, Column-Full and Column-Move), hard-blocks are fixed in columns and they are not free to move and the placer can move only logic-blocks which eventually leads to higher placement costs and larger channel widths. In case of BMR, however, both HBs and CLBs can be moved which increases the flexibility of the architecture; hence smaller placement cost and smaller channel widths.

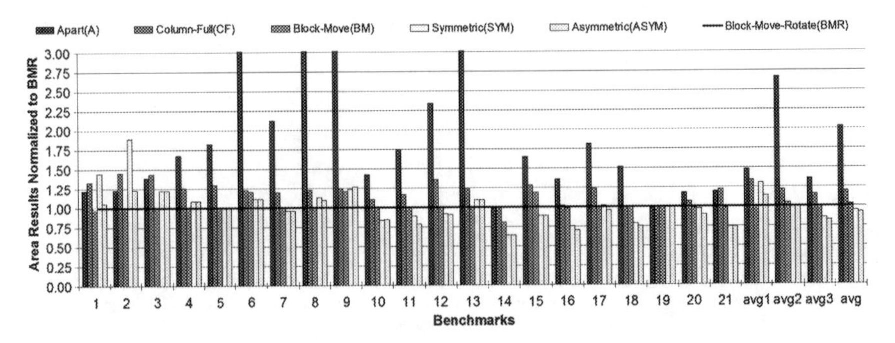

Fig. 4.15 Area comparison between different techniques of mesh-based and tree-based architecture

It is interesting to note that the only difference between BM and BMR techniques is that of block rotation. Although block rotation has no effect on CLBs, it affects the bounding box cost of HBs which can decrease the overall placement cost; hence a small gain in channel width is observed.

4.6.3.2 Area Comparison Results

Area results for different exploration techniques of mesh-based and tree-based architectures are shown in Fig. 4.15. Similar to the results of Figs. 4.13 and 4.14, area results of different techniques are normalized against the BMR technique of mesh-based architecture. In order to avoid congestion, results for only 4 out of 6 techniques of mesh-based architecture are presented here. It can be seen from the figure that gains of BMR technique observed in Figs. 4.13 and 4.14 remain valid here too and BMR technique gives either equal or better results when compared to other floor-planning techniques of mesh-based architecture. Among four floor-planning techniques of mesh-based architecture, on average, Apart gives the worst results and BMR gives the best results whereas the results of CF are in between Apart and BMR. However, when compared to exploration techniques of tree-based architecture, for SET I benchmark circuits, SYM requires 35% more area than BMR, and ASYM requires 12% more area than BMR. However for SET II benchmark circuits, on average BMR is almost equal to SYM and ASYM. For SET III benchmark circuits BMR is worse than SYM and ASYM by 14 and 18% respectively. Further, the comparison of ASYM with CF shows that for three sets of benchmarks, on average, tree-based architecture consumes 15, 21 and 29% less area than mesh-based architecture. As compared to the mesh-based architecture, tree-based architecture slightly under utilizes its logic resources because of its hierarchy. This under utilization leads to natural congestion spreading in routing resources; hence leading to smaller switch sizes and ultimately reduced area.

In this work, the BMR floor-planning serves as a near ideal floor-planning with which other floor-planning techniques are compared. It can be noted that results of CF compared to BMR vary depending upon the set of benchmarks that are used. For SET I benchmark circuits, where the types of blocks for each benchmark are two or more than two and communication is dominated by HB-HB type of communication, CF produces worse results than the other two sets of benchmarks. This is because of the fact that columns of different HBs are separated by columns of CLBs and HBs need extra routing resources to communicate with other HBs. However in BMR there is no such limitation; HBs communicating with each other can always be placed close to each other. For other two sets the gap between CF and BMR is relatively less. The reduced HB-HB communication in SET II and SET III benchmark circuits is the major cause of reduction in the gap between CF and BMR. However 21 and 16% area difference for SET II and SET III is due to the placement algorithm. In CF, the simulated annealing placement algorithm is restricted to place hard-block instances of a netlist at predefined positions. This restriction for the placer reduces the quality of placement solution. Decreased placement quality requires more routing resources to route the netlist; thus more area is required. The results show that BMR technique produces least placement cost, the smallest channel width and hence the smallest area for mesh-based heterogeneous FPGA. However, BMR floor-planning technique is dependant upon target netlists to be mapped upon FPGA. Although such an approach is not suitable for generalized FPGAs, it can be beneficial for domain specific FPGAs. Moreover, the hardware layout of BMR requires more efforts than CF.

For tree-based FPGA, ASYM produces better results than SYM because of its better logic resource utilization and further it is better than the best technique of mesh-based FPGA (i.e. BMR) by an average of 8% for a total of 21 benchmark circuits. The gain of tree-based FPGA is not large when compared to the best technique of mesh-based FPGA. This is because of the fact that tree-based FPGA requires more resources because of its hierarchy. Although it helps in spreading the congestion, it leads to extra logic and routing resources which decrease the area gain of tree-based FPGA. However, when the best technique of tree-based FPGA is compared to equivalent mesh-based FPGA (i.e. CF), it gives on average a gain of 15, 21 and 29% for SET I, SET II and SET III benchmarks respectively.

4.6.3.3 Critical Path Comparison Results

In order to evaluate the performance of different techniques of two architectures, we have calculated the number of switches crossed by critical path. Since we are exploring a number of techniques for both mesh-based and tree-based architectures, it would be very difficult to perform layout for each technique and determine the exact critical path delay. So, we use a simple model that gives an overview of the impact of active routing resources (switches) on the overall performance of the architecture. Similar to area results, critical path results are normalized against BMR floor-planning of mesh-based FPGA. These results are shown in Fig. 4.16. To avoid congestion, results for only 6 out of 8 techniques are shown. It can be seen from Fig. 4.16 that due to its higher flexibility, BMR gives higher performance results than other floor-planning techniques of mesh-based FPGA. On average, CF critical path crosses 5, 7 and 10% more switches than BMR technique for SET I, SET II and SET III bench-

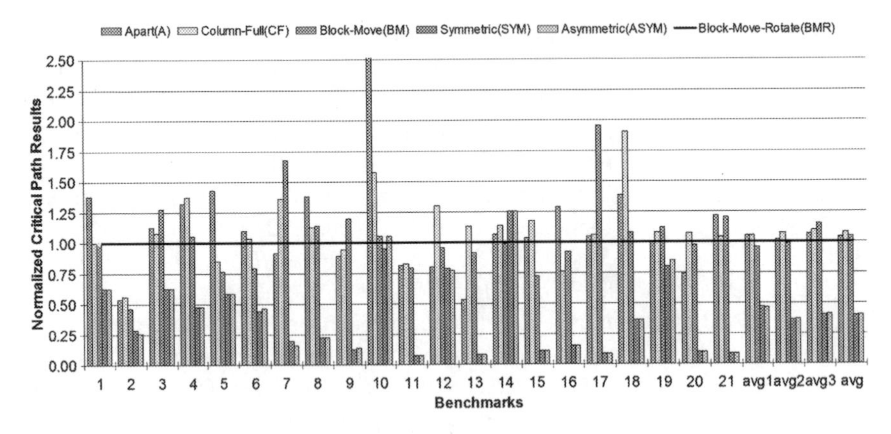

Fig. 4.16 Critical path comparison between different techniques of mesh-based and tree-based architecture

marks respectively. Although Apart (A) gives worst results in terms of placement cost, channel width and area, it is quite interesting to note that critical path results of Apart (A) are comparatively better than CF. This is because of the fact that in Apart columns of HBs are placed close to each other and apart from the CLBs. Since in SET I benchmarks majority of the communication involves HB-HB communication, there is a strong probability that critical path involves HBs which ultimately leads to 50% of benchmarks of SET I crossing less number of switches than CF. For SET II benchmarks this percentage drops to 44% as there is more communication between CLBs and HBs. However, in case of SET III benchmarks 75% of benchmarks cross less number of switches for Apart than CF as the communication pattern is dominated by CLB-CLB and in case of Apart there are no columns of HBs interspersed in between CLBs; hence leading to smaller number of switches that are crossed by critical path. Although Apart gives better results than CF, BMR manages to produce the best overall results among floor-planning techniques of mesh-based architecture due to its higher flexibility.

However, compared to the tree-based FPGA, both SYM and ASYM techniques of tree-based FPGA produce far better performance than BMR technique due to the inherent characteristic of tree-based architecture. Compared to BMR technique of mesh-based architecture, on average, SYM and ASYM techniques of tree-based architecture cross 53%, (54% less switches for SET I), 64%, (63% less switches for SET II) and 60%, (59% less switches for SET III) benchmarks respectively. It can also be observed from these results that on average ASYM technique crosses 1% more switches than SYM technique. In ASYM technique, HBs have a separate substructure and if critical path involves HBs and CLBs then it can lead to an increase in number of switches crossed by critical path (refer to Figs. 4.10 and 4.11). However if critical path involves no HBs or only HBs and I/Os, it can lead to smaller number of switches than SYM technique (refer to Fig. 4.16 results for benchmark 2 and 7).

4.6.3.4 SRAM and Buffer Comparison Results

Power optimization of FPGAs has become very important with the advancement in process technology. Although in this work a detailed power analysis of mesh-based and tree-based FPGA architectures is not performed, it gives a brief overview of the static power consumption of the two architectures; which has become increasingly important for smaller process technologies [12]. Static power of the FPGAs is directly related to the configuration memory and the number of buffers in an FPGA architecture [132]. Therefore, a comparison of configuration memory and number of buffers for different techniques of the two architectures is shown in Figs. 4.17 and 4.18 respectively.

Figure 4.17 shows number of SRAMs for different techniques normalized against the BMR technique of mesh-based FPGA. Comparison of BMR with CF shows that, on average, CF consumes 23, 16 and 9% more SRAMs than BMR for SET I, SET II and SET III respectively. Comparison of BMR with tree-based architecture techniques shows that, on average, SYM consumes 9% more and ASYM consumes 10% less SRAMs for SET I. However, for SET II and SET III SYM and ASYM consume 11%, 7% and 13%, 15% less SRAMs than BMR respectively. Similarly Fig. 4.18 shows that, compared to BMR, CF consumes 9, 22 and 18% more buffers for SET I, SET II and SET III respectively. Comparison of SYM and ASYM with BMR shows that both consume 6% more buffers for SET I, 3% less buffers for SET II and 15%, (18% less buffers for SET III). Although the comparison presented in Figs. 4.17 and 4.18 does not give detailed power estimation of the two architectures, it gives an empirical estimate of the static power of the two architectures and as stated by [63], it closely correlates to the average area results of the two architectures.

4.6.4 Results Using Generalized Experimentation Approach

In this approach experiments are performed for three sets of benchmarks shown in Tables 4.5, 4.6 and 4.7 respectively. Different exploration techniques are explored for each architecture. For each technique, a minimum common architecture is defined for each set of benchmarks that can implement any of the netlists of the benchmark set. Area comparison results for different exploration techniques of both mesh-based and tree-based architectures are shown Fig. 4.19. For each of the three sets of benchmark, the results of five floor-planning techniques of mesh-based architecture and two exploration techniques of tree-based architecture are normalized against the BMR technique of mesh-based architecture. Contrary to the individual experimentation approach where separate architectures are defined and optimized for each netlist of benchmark set, a common architecture is defined for each set in this approach.

The area comparison results of different floor-planning techniques show that compared to other floor-planning techniques of mesh-based FPGA, BMR produces equal or better results. However, compared to individual experimentation approach, the gain of BMR compared to CF is reduced from 23%, 10 to 5%, 3% for SET II and SET III

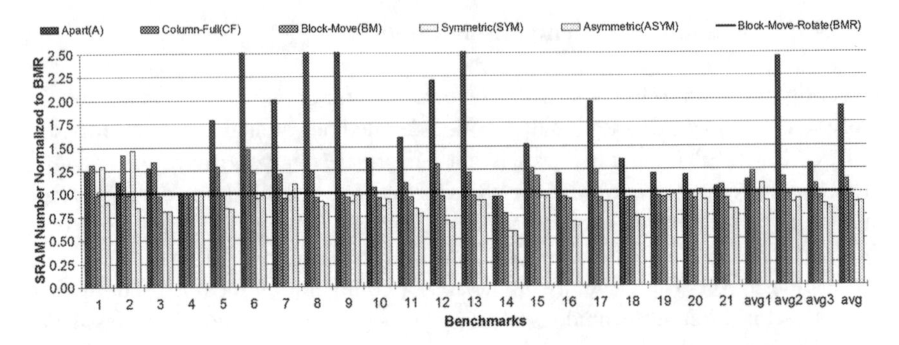

Fig. 4.17 SRAM comparison between different techniques of mesh-based and tree-based architecture

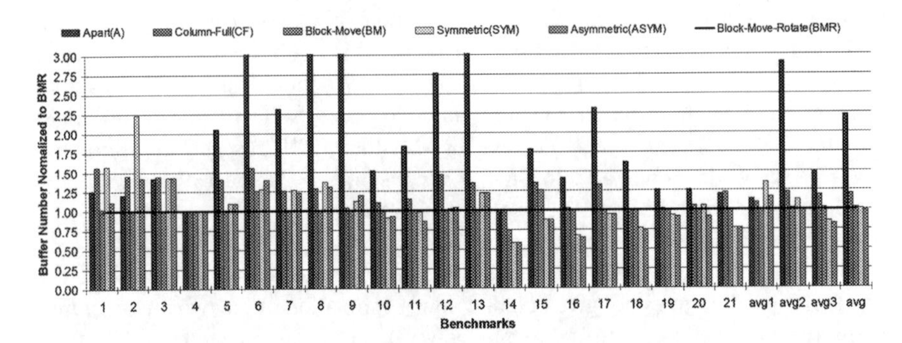

Fig. 4.18 Buffer comparison between different techniques of mesh-based and tree-based architecture

benchmarks respectively, while the gain for SET I benchmarks remains unchanged. This drop in gain is mainly due to the combined floor-planning optimization of all the netlists of a SET where the routing requirements of smaller netlists are overshadowed by those of larger netlists. As far as the comparison of BMR with SYM and ASYM techniques of tree-based FPGA is concerned, the results of tree-based topologies are further improved. For SET I benchmarks, SYM and ASYM techniques are only 4 and 3% worse than BMR and for SET II, their gain is increased from 0 to 22 and 24% and for SET III their gain is increased from 14–18% to 22–24% respectively.

In case of generalized experimentation, increase in the area gain of both techniques of tree-based architecture is because of the fact that in this experimentation the netlist having the highest routing requirements decides the logic and routing resources of the architecture. Routing requirement of a benchmark not only depends on the number of CLBs and HBs but also on the connection density. In our case the benchmarks with the highest routing requirements are benchmarks 4, 10 and 21 for SET I, SET II and SET III respectively and results of these benchmarks in Fig. 4.15 correspond well to the results of Fig. 4.19.

Generalized critical path comparison results for different techniques of mesh-based and tree-based architectures are presented in Fig. 4.20. It can be seen from the

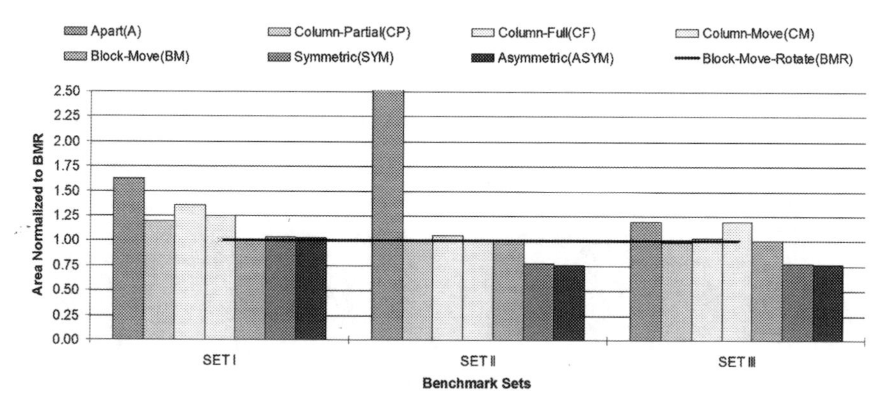

Fig. 4.19 Generalized area comparison between different techniques of mesh-based and tree-based architecture

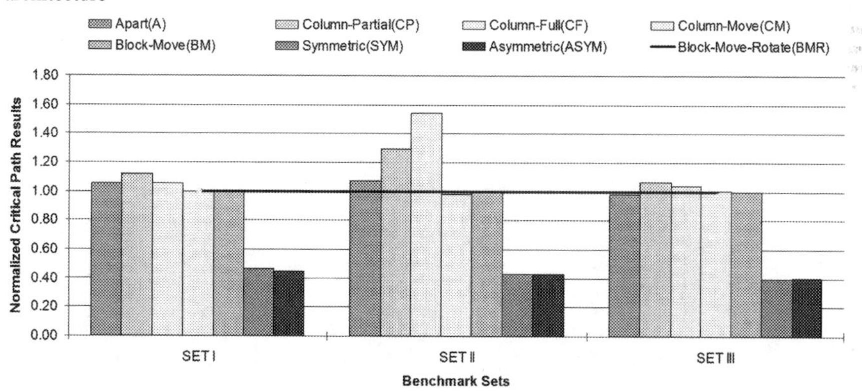

Fig. 4.20 Generalized critical path comparison between different techniques of mesh-based and tree-based architecture

figure that similar to results of individual experimentation methodology, techniques of tree-based architecture give far better results compared to the techniques of mesh-base architecture. On average, SYM crosses 54, 57 and 60% less switches than BMR and ASYM crosses 56, 57 and 60% less switches than BMR for SET I, SET II and SET III benchmark sets respectively. Further, SRAM and buffer comparison results are shown in Figs. 4.21 and 4.22 respectively. These results give an empirical estimate of the static power consumption of the two architectures which has become increasingly important with the advancement in the process technology. Figure 4.21 shows that CF consumes 37, 8 and 3% more SRAMs than BMR technique for SET I, SET II and SET III respectively whereas ASYM consumes almost same number of SRAMs for SET I and 24%, 14% less SRAMs than BMR for SET II and SET III respectively. Similarly the buffer comparison shows that ASYM produces the best overall results and as stated by [63] these results are in compliance with the area results of the two architectures. In this chapter we have mainly emphasized on the comparison between heterogeneous mesh-based and tree-based FPGA architectures

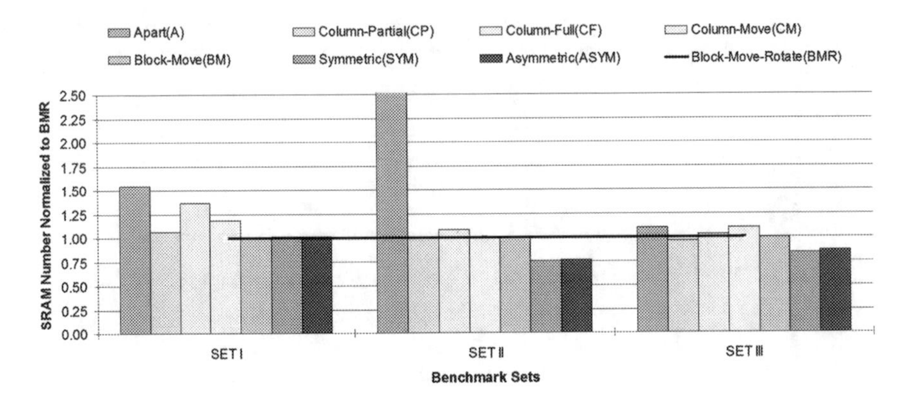

Fig. 4.21 Generalized SRAM comparison between different techniques of mesh-based and tree-based architecture

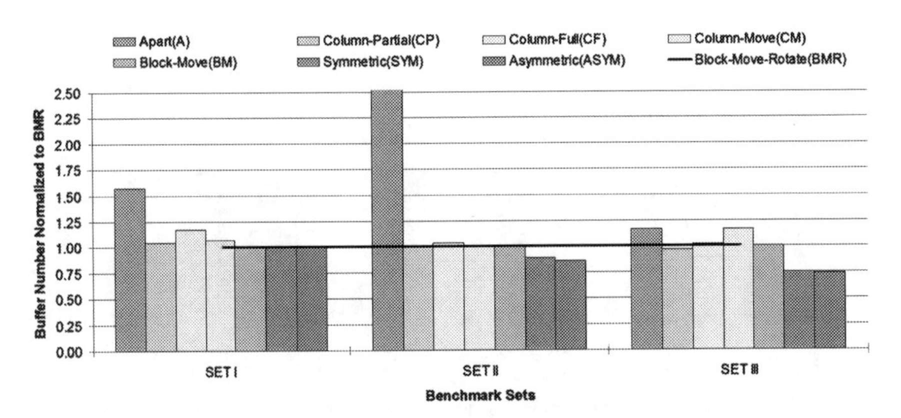

Fig. 4.22 Generalized buffer comparison between different techniques of mesh-based and tree-based architecture

and no comparison is presented between tree-based heterogeneous architectures and their homogeneous counterparts. However, in [45] we have performed a comparison between heterogeneous and homogeneous tree-based FPGA architectures and results show that on average heterogeneous architecture gives 41% area gain as compared to tree-based homogeneous FPGA architecture. So, it can be stated here that the introduction of hard-blocks in tree-based architectures improves their density as compared to their homogeneous counterparts and further they give better overall results when they are compared with mesh-based heterogeneous architectures.

The experimental results presented from Figs. 4.13 to 4.22 are concluded below. These conclusions are valid both for individual and generalized experimentation unless otherwise specified.

- Among the six floor-planning techniques of mesh-based FPGA, "Apart (A)" technique gives the worst area and static power results while "Block-Move-Rotate (BMR)" gives the best overall results. The "Apart" floor-planning is not a suitable floor-planning; atleast not for the netlists used in this work. This floor-planning might be advantageous if

1. Control-path portion of a circuit implemented on CLBs is relatively small as compared to data-path portion of circuit implemented on hard-blocks.
2. Routing network of control and data-path sections of the FPGA architecture are optimized independently.

- Column based floor-planning (CF) of hard-blocks is advantageous for an optimized tile-based layout generation; the widths of hard-blocks placed in columns can be appropriately adjusted to optimize the layout area. However, column-based floor-planning is unable to decrease the placement cost as some other floor-plannings do. This difference in placement costs can sometimes result in as high as 35% difference in total area of FPGA.
- The floor-planning achieved through Block Move and Rotate (BMR) operation gives the least possible placement cost, and eventually least FPGA area as compared to other floor-planning techniques. However, such a floor-planning can be achieved only if the set of netlists are known in advance. Such can be a case if an application specific FPGA is desired for a product.
- Among the two exploration techniques of tree-based architecture, "Asymmetric(ASYM)" technique gives the best overall results (area, critical path delay and static power) because of its better resource utilization (see Sect. 4.5.1). When compared to the equivalent mesh-based floor-planning technique (i.e. CF floor-planning technique), ASYM gives better overall results. However, when compared to the best floor-planning technique of mesh-based architecture (i.e. BMR), ASYM gives either equal or better results except for the SET I benchmarks which makes tree-based architecture unsuitable for benchmarks involving excessive communication between HBs. However, this deficiency can be remedied by incorporating the architecture modifications that are suggested in Sect. 4.5.1.

4.7 Heterogeneous FPGA Hardware Generation

The hardware of heterogeneous FPGA is generated in a similar manner as that of homogeneous FPGA. However, modifications are performed to incorporate the effect of different types of blocks that are used by different netlists. Similar to the homogeneous FPGA, the VHDL model generator of heterogeneous FPGAs is integrated with their exploration environments and the parameters that are supported by the exploration environment are also supported by the VHDL generator.

The FPGA generation flow remains exactly the same except that now the block database contains details about a variety of blocks that are used by different netlist being mapped on the architecture. As far as the remaining steps involved in the VHDL model generation are concerned, they remain the same. The only difference between the VHDL generation of homogeneous and heterogeneous FPGAs is that of the support for hard-blocks.

4.8 Summary and Conclusion

This chapter presented a new exploration environment for tree-based heterogeneous FPGA architecture which remains relatively unexplored despite its attractive characteristics. Different architectural techniques of tree-based architecture are then explored using its exploration environment. Exploration of heterogenous tree-based FPGA architecture is mainly related to the architecture description and architecture optimization. For the architecture description, a detailed architecture description mechanism is designed to define a heterogeneous FPGA architecture. Once the architecture is defined, it is optimized using different parameters of architecture exploration environments. Also, this chapter presented an exploration environment for mesh-based heterogeneous FPGA architecture. Contrary to the existing environments of mesh-based architecture that use pre-determined floor-planning, the environment presented in this chapter automatically optimizes the floor-planning of the architecture. The major feature of the exploration environments of two FPGA architectures is that they are flexible and they can be used to explore different exploration techniques for two FPGA architectures. In order to evaluate the exploration environments of two architectures, a generalized software flow is designed which maps different applications on the two architectures separately. The software flow used to explore the two architectures is flexible in the sense that it can be used to implement different types of applications having different types of blocks. Since communication trends of applications play a very important role in the architecture evaluation, special care has been taken while selecting these applications. We have selected 21 benchmarks (applications) where we have covered different aspects of inter-block communication.

A number of techniques are explored for both architectures using their respective software flows and experimental results show that for 21 benchmarks, on average, a column-based (CF) island-style FPGA takes 19% more area, crosses 8% more switches on critical path, consumes 13% more memories and 20% more buffers than the best non-column (BMR) based island-style FPGA. These differences increase as the number of different types of hard-blocks increase in the FPGA architecture. However, these gains might decrease due to difficulties associated with layout of non-column based heterogeneous FPGA. Further, the comparison between different techniques of mesh-based (island-style) and tree-based (hierarchical) architecture shows that, on average, the best technique of tree-based architecture is 8.7% more area efficient, crosses 60% less switches on critical path, consumes 11% less memories and almost same number of buffers than the best non-column based technique of mesh-based architecture. These gains further increase when the best technique of tree-based architecture is compared to the equivalent column-based technique (i.e. CF) of mesh-based architecture. These results are averaged for 21 benchmarks which cover different aspects of heterogeneous benchmarks. The work and results presented in this chapter are also published in [116].

Chapter 5
Tree-Based Application Specific Inflexible FPGA

The FPGA architectural developments enabled by advancement in process technology have greatly enhanced their area efficiency, performance and power consumption. However, use of FPGAs as general purpose and field programmable devices leads to a large overhead and eventually makes them unsuitable for applications requiring high area density, superior performance or very small power consumption. In such a case, other devices like ASICs or structured-ASICs might be considered as an alternative. However, these alternatives have issues associated to either NRE cost, time to market or reconfigurability. In fact FPGAs provide a large degree of freedom in their architectures which has not been fully explored yet and there is a big room for improvement in FPGA architectures. In this chapter we present a method to optimize an FPGA for a particular domain of applications. Such an FPGA is named as Application Specific Inflexible FPGA (ASIF).

An Application Specific Inflexible FPGA (ASIF) is a modified form of FPGA with reduced flexibility and improved density. It is designed for a predetermined set of applications that operate at mutually exclusive times. In this chapter a new tree-based ASIF is presented and four ASIF generation techniques are explored for a set of 16 MCNC benchmarks. A comparison between different tree-based ASIFs and an equivalent tree-based FPGA is also presented. Later, for tree-based ASIF, the effect of lookup table and arity size is explored for the most efficient technique among the four explored techniques. Further a detailed comparison between tree-based ASIF and an equivalent mesh-based ASIF is performed. Finally a quality analysis of tree-based ASIF and quality comparison between mesh-based and tree-based ASIFs is performed.

5.1 Introduction and Previous Work

Medium to low volume production of FPGA-based systems is quite effective and economical because FPGAs are easy to design and program in shortest possible time. The generic reconfigurable resources of FPGAs lead to larger area, poor performance

U. Farooq et al., *Tree-Based Heterogeneous FPGA Architectures*,
DOI: 10.1007/978-1-4614-3594-5_5,

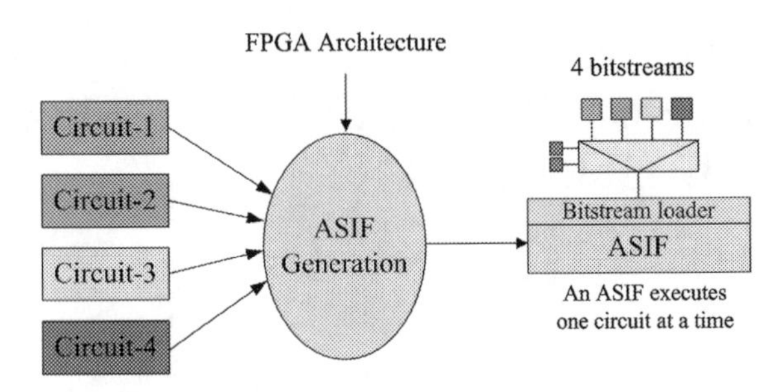

Fig. 5.1 An illustration of ASIF generation concept [94]

and higher power consumption as compared to ASIC. As a result of these drawbacks, FPGAs become unsuitable for some applications. To address this limitation a range of alternatives exists.

The primary alternative to an FPGA is an ASIC that has speed, power and area advantages over an FPGA. However, ASIC design requires huge resources in terms of time and money and has become increasingly complicated with advancement in process technology. The difficulties associated with the design process of ASICs have led to the development of Structured-ASICs. Structured-ASICs can cut the NRE cost of ASICs by more than 90% while speeding up significantly their time to market [125]. Structured-ASICs are explored or manufactured by several companies [41, 91, 103, 125].

FPGA vendors have also started giving provision to migrate FPGA based application to Structured-ASIC. In this regard, Altera has proposed a clean migration methodology [98] that ensures equivalence verification between FPGA and its Structured-ASIC (known as HardCopy [56]). However, migration of an FPGA based application to HardCopy can execute only a single circuit and it totally loses the quality of reconfigurability. An ASIF, on the other hand, comprises of optimized logic and routing resources like Structured-ASIC but retains enough flexibility to implement a set of pre-determined applications that operate at mutually exclusive times. Figure 5.1 illustrates the ASIF generation concept. In the final phase of development cycle of an FPGA based product, if the set of circuits to be mapped on the FPGA are known, it can be reduced to an ASIF for all the given set of circuits. An ASIF can give considerable area, performance and power gains to an FPGA-based product by reducing it to a much smaller multiplexed circuit. Execution of different application circuits can be switched by loading their respective bitstream on ASIF.

The concept of an ASIF is similar to configurable ASIC (cASIC) cores [35]. cASIC is a reconfigurable device that can implement a given set of circuits which operate at mutually exclusive times. However, cASIC and ASIF have several major differences. cASIC is mainly intended as an accelerator in a domain-specific systems on chip, and it supports only full-word logic blocks (such as 16-bit wide multipliers, adders, RAMs etc.) to implement data-path circuits. However, an ASIF supports both soft-blocks and hard-blocks and it can be used to implement an entire design. Similarly, cASIC uses segmented bus-based routing where signals are routed in 16-bit wide buses whereas ASIF uses fine-grain routing where signals are routed in a bit by bit manner.

In [93], a mesh-based ASIF is initially presented. In this chapter, we present a new tree-based ASIF and compare it with mesh-based ASIF. This chapter mainly concentrates on the exploration and comparison of different ASIF generation techniques for tree-based architecture using a set of 16 MCNC benchmarks. A comparison between mesh-based and tree-based ASIFs is performed later and finally the chapter is concluded with the quality analysis of the two architectures. In this chapter the architectures with only homogeneous blocks are considered. Experimentation for heterogeneous architectures is presented in the next chapter.

5.2 Reference FPGA Architectures

This section gives a brief overview of mesh-based and tree-based FPGA architectures. Application circuits are efficiently placed and routed on these architectures and later they are reduced to their respective ASIFs.

5.2.1 Reference Tree-Based FPGA Architecture

A tree-based architecture is a hierarchical architecture having unidirectional interconnect [132]. Tree-based architecture exploits the locality of connections that is inherent in most of the application designs. In this architecture, CLBs and I/Os are partitioned into a multilevel clustered structure where each cluster contains sub-clusters and switch blocks allow to connect external signals to sub-clusters. The number of signals entering into and leaving from the cluster can be varied depending upon the netlist requirement. However, they are kept uniform over all the clusters of a level. The bandwidth of clusters is controlled using Rent's rule [74] which is easily adapted to tree based architecture. This rule states that

$$IO = k \cdot n^{\ell \cdot p} \tag{5.1}$$

where ℓ is a tree level, n is the arity size, k is the number of in/out pins of a LUT and IO is the number of in/out pins of a cluster at level ℓ. p is the factor which controls the cluster bandwidth at each level of the tree-based architecture. Further details about tree-based architecture are already explained in Chap. 3.

5.2.2 Reference Mesh-Based FPGA Architecture

The reference mesh-based FPGA is a Versatile Place and Route (VPR)-style [22] architecture. A mesh-based architecture contains Configurable Logic Blocks (CLBs) that are arranged on a two dimensional grid. Each CLB contains one Look-Up table with 4 inputs and 1 output (LUT-4), and one Flip-Flop (FF). Further architectural details of mesh-based homogeneous architecture are already presented in Chap. 3.

5.3 Software Flow

The detail of the software flow used to place and route different netlists (benchmarks) on the two architectures is shown in Fig. 3.9. The flow starts with the technology independent optimization of the circuit [102]. The mapping of the circuit is then performed which converts logic expressions into LUTs of a given size (K). After mapping, packing is performed using T-VPACK [14] that packs registers together with K-input LUTs and converts the netlist into net format. A netlist in net format contains CLBs and I/O instances that are connected together using nets. The size of a CLB is defined as the number of LUTs contained in it and in this work this size is set to be 1 for both mesh-based and tree-based architectures. Once netlist is obtained in net format, it is placed and routed separately on tree-based and mesh-based FPGA. Further detail regarding the placement and routing of different benchmarks on the two architectures is already given in Sect. 3.4.

5.4 ASIF Generation Techniques

Reconfigurability of FPGAs is their biggest asset but at the same time it is also their largest drawback as it makes them larger, slower and more power consuming. Customized reconfigurable architectures like ASIF can reduce these overheads of FPGA while maintaining a certain degree of flexibility. An ASIF is reduced from an FPGA where a set of predetermined applications are placed and routed and later all unused routing reconfigurability is removed to generate ASIF. In order to generate an ASIF, first, a minimal common FPGA architecture is defined that can implement any application of the set. Netlists are then individually placed and routed on the architecture and later all unused resources are removed to generate an ASIF.

ASIF generation concept can be understood with the help of Fig. 5.2. In Fig. 5.2a, a simple netlist is mapped on a tree-based FPGA architecture. It is a two level architecture that can map any application having 2 inputs, 2 outputs, and 4 logic blocks each of which has 1 input and 1 output. In this architecture, logic blocks are connected to each other through downward and upward switch blocks. In this figure "DMSB" is a downward switch block and "UMSB" is an upward switch block. These switch blocks are full cross bars and are used to connect different wires of the architecture. The netlist mapped on the architecture comprises of 2 main inputs, 2 main outputs, and 4 blocks. The 2 inputs are connected to the inputs of block "A", "C" and 2 outputs are connected to the outputs of block "B", "D" respectively. Further the output of block "A" is connected to block "B" and output of block "C" is connected to block "D". In this figure the wires used by the netlist after placement

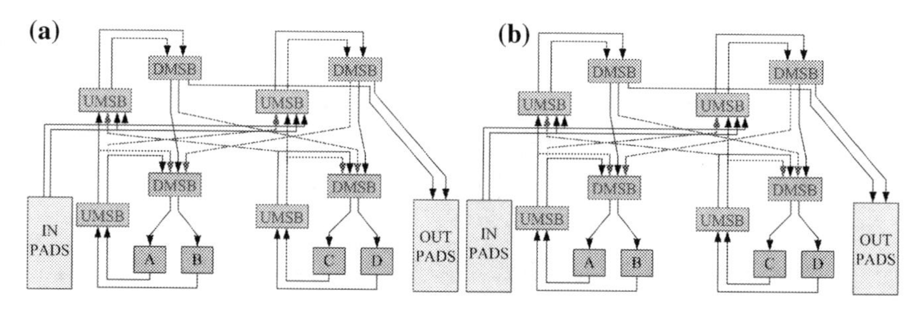

Fig. 5.2 ASIF generation explained with a simple example. **a** Netlist on FPGA; **b** netlist on ASIF

and routing are highlighted in black while unused resources are kept in blue. As it can be seen from the figure that only a handful of total available resources are used and rest of them go unused. This is the price that we pay for being general purpose and reconfigurable. Figure 5.2b shows the ASIF generated for this netlist where switch blocks are replaced with simple wire connections and all unused routing resources are removed. Although in this example only a single netlist tree-based ASIF generation is explained, an ASIF can be generated for as many netlists as we wish where enough flexibility is retained to implement any of those netlist while removing unused resources. Since an ASIF is designed for a specific set of applications, it can not implement any application out of that set.

Because of the generalized nature of FPGAs, conventional placement/partitioning, routing algorithms used for FPGAs are designed for individual netlists and they do not consider the inter-netlist dependence. Since an ASIF is designed for a group of predetermined applications, inter-netlist dependence should be taken into account to optimize the use of logic and routing resources of the architecture. So these algorithms are modified to efficiently share the logic and routing resources of the architecture. The modifications made in the placement/partitioning and routing algorithms have led to four different ASIF generation techniques that are explained in the remaining part of the section. These techniques are general in nature and they are applicable to both mesh-based and tree-based architectures unless otherwise specified.

5.4.1 ASIF-Normal Partitioning/Placement Normal Routing

This technique is the simplest among the four techniques. Once the FPGA architecture is defined for a predetermined set of netlists, netlists are placed and routed individually on the architecture. In this technique the partitioning/placement and routing algorithms remain unchanged and the dependance between different netlists is not taken into account. However, provisions are made to allow the placement/partitioning and routing of multiple netlists on the architecture in a manner similar to [35]. The CLBs, I/O blocks on the architecture, and the routing resources are shared by all

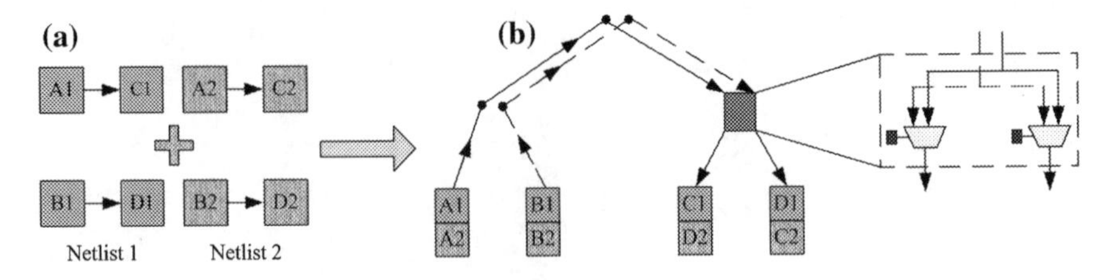

Fig. 5.3 **a** Two simple netlists; **b** a tree-based ASIF generated using ASIF-NPNR technique

netlists. So the target FPGA architecture contains maximum number of CLBs and maximum signal bandwidth required by any of the netlist in the group of netlists.

In order to place multiple netlists on the architecture, mapping of instances of different netlists is allowed on each block position of the architecture. However, mapping of multiple instances of same netlists is not allowed on a single block position. Similarly for routing multiple netlists on the architecture, sharing of routing wires among nets of different netlists is allowed. But the sharing of routing wires among nets of same netlists is not allowed. Once the placement and routing of all the netlists is over, unused switches are removed from the architecture. The retained switches belong to different switch blocks of the architecture that have enough flexibility to implement any of the netlists of the group. A simple example of ASIF generation using ASIF-normal partitioning/placement normal routing (ASIF-NPNR) technique is shown in Fig. 5.3. In this figure an ASIF is generated for two simple netlists and it requires only one switch. Since an ASIF is generated for a group of predetermined netlists that share a common architecture, the inefficient use of the resources of the common architecture may undermine the overall efficiency of the architecture. Although this technique may give better results when compared to FPGA as it has reduced number of switches and associated configuration memory, it gives poor results when compared to ASIF generation techniques that efficiently use logic and routing resources of the common architecture. For example, the switch used by this technique in Fig. 5.3 could have been avoided by efficiently sharing the logic resources of the architecture and the resulting architecture would have required no switches at all (for details see Sect. 5.4.2).

5.4.2 ASIF-Efficient Partitioning/Placement Normal Routing

In this technique the netlists are efficiently partitioned/placed on the architecture while routing of the netlists is performed in a manner similar to the previous technique. An FPGA architecture basically comprises of the logic and routing sources and the main objective of this technique is to share the logic sources of the architecture among different netlists in such a way that overall area of the architecture is reduced. The main idea behind the inter-netlist logic optimization (efficient logic sharing) is to have common source and destination positions among instances of

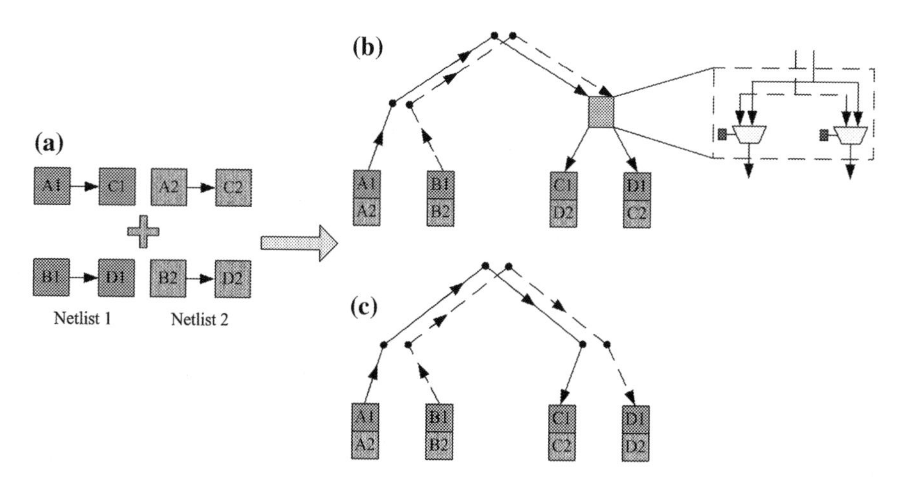

Fig. 5.4 **a** Two simple netlists; **b** ASIF generated using inefficient logic sharing; **c** ASIF generated using efficient logic sharing

different netlists of the group. Inter-netlist optimization can be understood with the help of Fig. 5.4. Figure 5.4a shows two simple netlists, Fig. 5.4b shows mapping of two netlists on a simple tree-based ASIF where inter-netlist dependance is not exploited which leads to the inclusion of unnecessary switches in the architecture. Figure 5.4c shows the efficient mapping of two netlists on the architecture which can eventually lead to fewer number of switches compared to the inefficient mapping. The efficient logic sharing is performed by considering inter-netlist optimization while not compromising the intra-netlist optimization.

For tree-based ASIF, efficient partitioning (logic sharing) is achieved by simultaneously partitioning multiple netlists. The main idea behind the efficient partitioning is to have a certain percentage of common source and destination positions across the group of netlists under consideration. This objective is achieved by fixing a percentage of block positions and then propagating these positions through all the netlists of the group. Through this technique we manage to have netlists with common source and destination positions. Although, fixing the block positions may lead to increased number of common source and destination positions across a number of netlists; hence improved inter-netlist logic sharing is achieved. But this can compromise the intra-netlist partitioning optimization which leads to poor intra-netlist partitioning and hence making the resulting architecture more resource hungry. In other words, inter-netlist and intra-netlist optimizations are linked to each other in an inverse manner and care must be taken to find a trade-off between the two optimizations which may lead to an overall optimized area of the architecture. In our case 15, 85 combination gives the best area results where 15% instances of each netlist are reserved to serve the inter-netlist optimization purpose while 85% instances are at the disposal of partitioner so that it can perform the intra-netlist optimization. Although this combination gives good results when compared to non-efficient partitioning, considerable room exists to improve this approach (For details refer to Sect. 5.6.4).

$$\begin{aligned}
&\text{Cost} = ((\text{W} * \text{BBX}) + ((100 - \text{W}) * \text{DC} * \text{Normalization Factor}))/100 \\
&\text{Where } 0 \leq \text{W} \leq 100 \\
&\text{Normalization Factor} = \text{Initial BBX/Initial DC} \\
&\text{BBX} = \text{Bounding Box Cost} \\
&\text{DC} = \text{Driver Count Cost}
\end{aligned} \tag{5.2}$$

For mesh-based ASIF, the efficient logic sharing is performed using Eq. 5.2. This equation gives the cost function of placement algorithm where intra-netlist placement is optimized by minimizing BBX cost; inter-netlist placement is optimized by minimizing Driver Count (DC) cost. DC cost is the sum of the driver blocks targeting the receiver blocks of the architecture over all the netlists of the group. If more driver instances of different netlists share a common position on the architecture and their respective receiver instances also share a common position then DC cost is said to be small and vice-versa. As DC cost and BBX cost are not of the same magnitude, they are made comparable by using a normalization factor. The influence of the two costs on the total cost of the function is controlled by factor "W". For our experimentation the value of "W" is set to be 80 because it gives best area results.

5.4.3 ASIF-Normal Partitioning/Placement Efficient Routing

In this method, the netlists are placed separately on the FPGA and efficient logic sharing is not performed. But routing is done efficiently in order to minimize the required number of switches and routing wires. This is done by maximizing the shared switches required for routing all the netlists on the FPGA. The efficient wire sharing encourages different netlists to route their nets on an FPGA with maximum common routing paths. After all the netlists are efficiently routed on FPGA, unused switches are removed from the architecture to generate an ASIF.

The pathfinder routing algorithm is modified to perform efficient wire sharing for the group of netlists. Netlists are routed in a sequence. Each routed netlist saves information that which nodes and edges it has used. Later, the next netlist uses this information to perform efficient routing. Figure 5.5 explains different routing scenarios when 2 netlists are routed on a routing graph. The graphical representation in Fig. 5.5a shows a case in which two nodes occupied by nets of 2 different netlists drive the same node. Figure 5.5b shows a case in which nodes occupied by nets of different netlists use different edges to drive different nodes. In Fig. 5.5c, both netlists share same node and edge to drive same node. Finally, Fig. 5.5d shows a node shared by both netlists drives different nodes. The physical representation for each of the four sub graphs are also presented along. In order to reduce the total number of switches and total wire requirement, the physical representation in Fig. 5.5 suggests that case (a) must be avoided because it increases the number of switches (here a

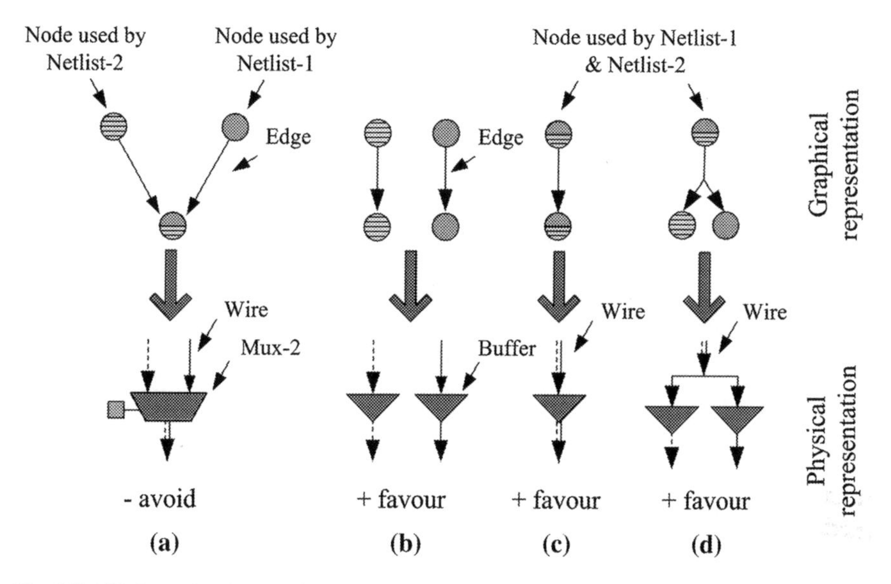

Fig. 5.5 Efficient wire sharing [94]. **a** Nodes use different edges to drive same node; **b** nodes use different edges to drive different nodes; **c** node uses the same edge to drive same node; **d** node uses the different edges to drive different nodes

mux-2), whereas case (b–d) should be favored. Favoring these cases means that if more routing resources exist in FPGA architecture, it is more probable to exploit such cases. For this reason, in order to create more routing resources, Sect. 5.6 performs experiments with varying channel widths.

The routing preferences shown in Fig. 5.5 need to be integrated in the pathfinder routing algorithm. For this purpose, the cost function of a node is modified in a similar way as for timing-driven cost function [22]. A particular routing is avoided or favored by increasing or decreasing the cost of a node. If a net is to be routed from current node to next node, the cost of next node can be calculated with the formulas shown in Eq. 5.3. The cost of a node depends on its congestion cost (for details refer to Eq. 2.3). Here, the increase or decrease in cost is controlled by a constant "Factor". The value of this factor ranges between 0 and 0.99. If an FPGA architecture has limited routing resources, a maximum value of factor might not allow the routing algorithm to resolve all congestion problems. So the value of factor is gradually decreased if the routing solution does not converge after a few routing iterations.

$$\begin{aligned} &\text{(Normal)} \quad \text{Cost}(n) = \text{Congestion Cost}(n) \\ &\text{(Avoid)} \quad \text{Cost}(n) = (1 + \text{Factor}) * \text{Congestion Cost}(n) \\ &\text{(Prefer)} \quad \text{Cost}(n) = (1 - \text{Factor}) * \text{Congestion Cost}(n) \\ &\text{Where } 0 \le \text{Factor} \le 0.99 \end{aligned} \tag{5.3}$$

Table 5.1 Area of different cells

Block name	Inputs	Outputs	Block size (λ^2)
clb	4	1	58,500
sram	–	–	1,500
buffer	1	1	1,000
flip-flop	1	1	4,500
mux 2:1	2	1	1,750

Multiple netlists can be routed on an FPGA architecture sequentially. In sequential routing, each routed netlist saves the information of the routing path it has used. Later, the next netlist uses this information to perform efficient routing by giving preference to some nodes over others. Experiments are done with netlists sequenced in different orderings (i.e. netlists ordered in ascending or descending order according to their size, signal requirements etc). An ASIF generated with netlists routed sequentially in descending order of their signal bandwidth requirements gives minimum area results. However the area difference between ASIFs (generated using different netlist orderings) becomes negligible as the routing resources increase.

5.4.4 ASIF-Efficient Partitioning/Placement Efficient Routing

In this method we combine the advantages of Sects. 5.4.2, 5.4.3 where both partitioning and routing of the netlists are performed in an efficient manner (refer to Figs. 5.4, 5.5). Netlists are efficiently placed and routed on the architecture and later all the unused routing resources are removed to generate an ASIF. Effect of different ASIF generation techniques on the tree-based architecture is shown in the next section.

5.5 ASIF Area Model

A generic area model is used to calculate the area of different ASIFs. The area model is based on the reference FPGA architecture explained in Sect. 5.2. Area of SRAMs, multiplexors, buffers and Flip-Flops is taken from a symbolic standard cell library (SXLIB [9]) which works on unit Lambda(λ). Area of different cells used for the area calculation is shown in Table 5.1. When an ASIF is generated, all unused resources are removed. With the removal of switches, wires are connected with one another to form long wires. The area of ASIF is reported as the sum of the areas taken by multiplexors, SRAMs, buffers and CLBs. The area model also reports the total number of routing wires used for routing all netlists. In the next section the term "Routing area" is used for area taken by multiplexors, SRAMs and buffers. The term "Logic area" is used for area taken by "CLBs". Since experiments are performed for a varying range of signal bandwidths and an ASIF affects only routing area with changing signal bandwidth; few of the area comparisons in the next section are done only for "Routing area".

Table 5.2 Description of circuits used in experiments

Index	Circuit name	Number of inputs	Number of outputs	Number of 4-input LUTs
1	pdc	16	40	3,832
2	ex5p	8	63	982
3	spla	16	46	3,045
4	apex4	9	19	1,089
5	frisc	20	116	2,841
6	apex2	38	3	1,522
7	seq	41	35	1,455
8	misex3	14	14	1,198
9	elliptic	131	114	2,712
10	alu4	14	8	1,242
11	des	256	245	1,506
12	s298	4	6	1,091
13	bigkey	229	197	1,147
14	diffeq	64	39	1,161
15	dsip	229	197	1,145
16	tseng	52	122	953

5.6 Experimental Results and Analysis

This section is basically divided into three parts. First of all we explore the effect of different ASIF generation techniques on tree-based ASIF. For each ASIF generation technique, tree-based ASIF is generated for a set of 16 MCNC benchmarks. Details of these benchmarks are shown in Table 5.2. Results of ASIF generation techniques are then compared with those of tree-based FPGA. Secondly, experiments are performed to determine the effect of LUT and arity size on a tree-based ASIF. It is very important to determine the best LUT and arity size for a tree-based architecture because an inappropriate combination can undermine the overall efficiency of the architecture. Finally, a comparison between mesh-based and tree-based ASIFs is performed using their best ASIF generation techniques.

5.6.1 Effect of Different ASIF Generation Techniques on Tree-Based Architecture

Four ASIF generation techniques are explored for tree-based ASIF. For each technique an ASIF is generated for a group of 16 MCNC benchmarks shown in Table 5.2. In this table, benchmarks with LUT size 4 are shown as this is the size that we use initially for the comparison between different techniques and they are ordered according to their bandwidth requirement.

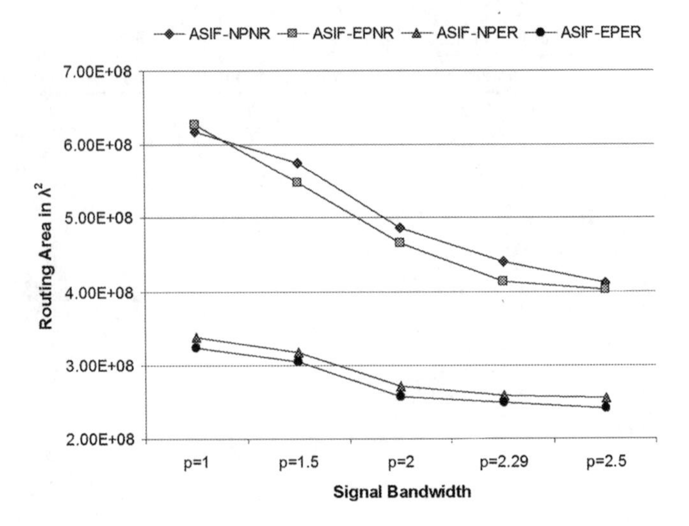

Fig. 5.6 Routing area comparison between different ASIF generation techniques

Figure 5.6 shows the variation in routing area of tree-based architecture using different ASIF generation techniques. These results are obtained using LUT-4, arity 4 combination while signal bandwidth is varied between a minimum and maximum value. Signal bandwidth values correspond to value of p in Fig. 5.6 and it is calculated using Eq. 5.1. Since a change in signal bandwidth does not affect the logic area of the architecture, only the variation in routing area with varying signal bandwidth is shown in Fig. 5.6. For each ASIF generation technique, signal bandwidth is varied from an initial minimum value which is required to place and route all the netlists of the group. The signal bandwidth is then increased gradually to a value beyond which a further increase in signal bandwidth does not improve the area of architecture. Although the rate of decrease in routing area of the architecture with increasing signal bandwidth varies, a decrease in the routing area of the architecture for each ASIF generation technique is observed. The difference in the routing areas of different ASIF generation techniques is mainly because of the different optimization techniques that are employed in each ASIF generation technique. The decrease in the routing area of the architecture with increasing signal bandwidth is because of the better use of cases that are shown in Fig. 5.5. Although the techniques with normal routing algorithm are not designed to exploit these cases, when there are abundant routing resources, routing algorithm automatically uses the less congested available resources that ultimately leads towards smaller routing areas with increased signal bandwidths. However, ASIF techniques with efficient routing algorithms exploit these cases more appropriately and this is the factor that contributes towards the larger area gap between techniques using efficient routing (i.e. ASIF-NPER, ASIF-EPER) and techniques not using efficient routing (i.e. ASIF-NPNR, ASIF-EPNR). As can be seen from the figure that ASIF generation technique with no optimization (i.e. ASIF-NPNR) produces the worst results among four ASIF generation techniques.

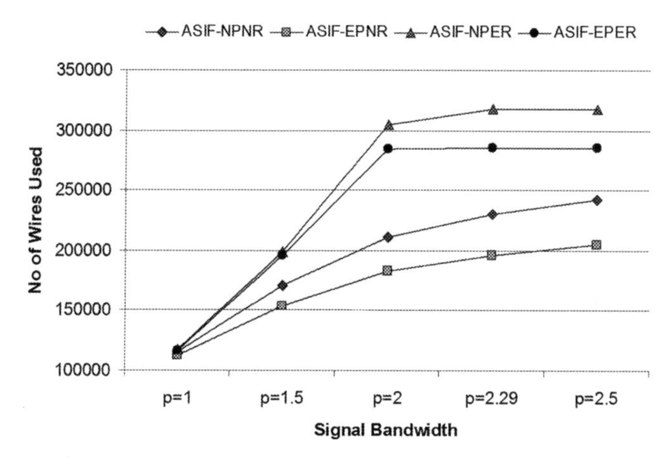

Fig. 5.7 Wire comparison between different ASIF generation techniques

Also the technique with only efficient partitioning solution (i.e. ASIF-EPNR) is not as efficient as the technique with only efficient routing solution (i.e. ASIF-NPER). Efficient partitioning alone is unable to produce any significant results because it is unable to exploit the routing resources of the architecture that contribute a major percentage of the total area. However, when it is combined with efficient routing, it gives the best overall results compared to the rest of ASIF generation techniques.

Figure 5.7 shows the variation in the number of wires that are used by different ASIF generation techniques. Similar to area results, signal bandwidth is varied from the minimum required to the maximum. Results show that ASIF-efficient partitioning/placement normal routing (ASIF-EPNR) uses the least number of wires while ASIF-normal partitioning/placement efficient routing (ASIF-NPER) uses the most number of wires. It can be noticed from the figure that techniques using efficient routing resource sharing consume more wires as compared to the techniques that are not using efficient routing resource sharing. This is because of the fact that when there are abundant routing resources in the architecture, techniques with efficient routing resource sharing tend to use less congested wires to avoid the increase in number of switches; hence causing an increase in the total number of wires. On the other hand, techniques that are not using efficient routing resource sharing do not hesitate to use the already congested wires which results in larger switches but smaller number of wires.

Further, it can be noticed from the figure that techniques with efficient logic resource sharing consume less number of wires than the techniques that are not using efficient logic sharing. This is because of the fact that when efficient logic sharing is used, more instances share a common source to destination path across the group of netlists hence resulting in less number of total wires being used. These two facts combined together result in the trends that are shown in Fig. 5.7.

Figure 5.8 shows the comparison of different ASIFs generation techniques where ASIFs are generated for a varying range of netlists and the order shown in Table 5.2 is

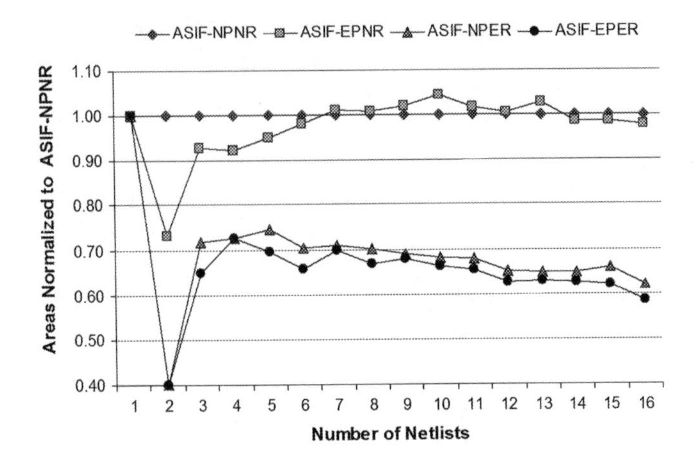

Fig. 5.8 Comparison of different ASIFs normalized to ASIF-NPNR

followed. In this figure "1" on the X-axis means that ASIF is generated for first netlist of Table 5.2, "2" means that ASIF is generated for first 2 netlists of the table and so on. For all four techniques, the signal bandwidth is set to be maximum that produces the best results. As it can be seen from the figure that ASIF-NPNR produces the worst results among 4 techniques and ASIF-EPNR produces slightly better results as compared to it. This is because of the efficient logic sharing employed in this technique. But the difference is not significant as the efficient logic sharing alone is of no use without efficient routing sharing. However, efficient routing resource sharing alone produces significant results and when combined with efficient logic sharing, it gives the best results with a gain of 41% for 16 benchmarks over ASIF-NPNR.

In order to perform the performance comparison between different ASIF generation techniques, we have implemented a simple model. This model calculates the number of switches that are crossed by critical path of different benchmarks being implemented on the ASIF. These critical path switch numbers are then averaged across all the benchmarks to determine the average number of critical path switches for a particular ASIF generation technique. For four ASIF generation techniques, average number of critical path switches for a varying range of signal bandwidths are shown in Fig. 5.9. Similar to area results, ASIF-efficient partitioning/placement efficient routing (ASIF-EPER) produces best critical path results because of its better resource utilization approach; hence leading to best performance results. Although the model that is employed to calculate critical path delay is empirical and it does not give us the exact delay values. However, this model gives us an idea about the effect of different ASIF generation techniques on the performance of tree-based architecture.

Figure 5.10 shows the comparison between tree-based FPGA and the worst and the best ASIF generation techniques. For ASIF generation techniques, the signal bandwidth is set to be maximum that produces best results and for tree-based FPGA a minimum signal bandwidth is determined that is required to implement any of

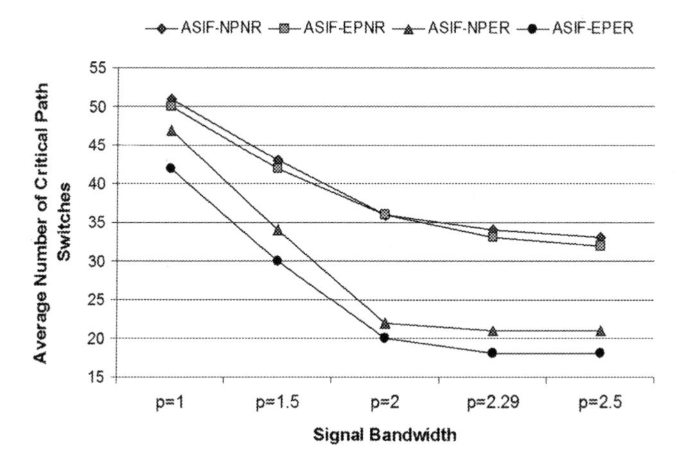

Fig. 5.9 Critical path comparison between different ASIF generation techniques

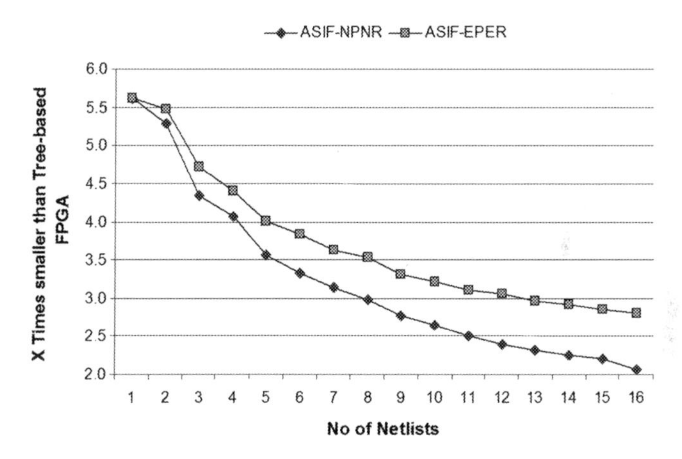

Fig. 5.10 Tree-based FPGA versus tree-based ASIFs

the netlists under consideration. In this figure, starting from 1, ASIFs are generated for a varying range of netlists and results are then compared to those of tree-based FPGA. When varying the number of netlists, the order shown in Table 5.2 is followed. For example, netlist "1" means an ASIF is generated for "pdc" only and netlist "2" means that an ASIF is generated for "pdc" and "ex5p" and so on. Results of ASIFs for varying number of netlists are compared against a LUT-4, arity-4 tree-based FPGA that can implement any of the netlists of the Table 5.2. It can be seen from the figure that for best ASIF generation technique, for 1 netlist ASIF-EPER is 5.6 times or 82% smaller than tree-based FPGA and as the number of netlists increase, the gap between the two decreases and for 16 netlists ASIF-EPER is 2.8 times or 64% smaller than the tree-based FPGA. The results in Fig 5.10 suggest that ASIF gives considerable area gain for a small group of netlist. However, the advantage of ASIF begins to fade as the number of netlists increase and for a very large group of netlists it might not be able to give any significant area advantages. So far in this section we have presented a comparison between different ASIF generation techniques for tree-based architecture. Results of the comparison suggest that among four techniques, ASIF-EPER gives best overall results and this is the technique that will be used as base for further experimentation.

5.6.2 *Effect of LUT and Arity Size on Tree-Based ASIF*

An appropriate combination of LUT and arity size plays a very important role in the efficiency of the architecture. Usually when LUT size (K) or arity size (N) is increased, the functionality of base-cluster of the architecture is also increased which reduces overall number of base-clusters that are required to implement a certain function. But on the other hand, area of base-cluster increases with increase in K or N. So an increase in K or N may increase base-cluster functionality and decrease the total number of base-clusters that are required to implement a certain benchmark. But it can leave a bad effect on the overall area of the architecture due to increased area of base-clusters. In this work we analyze the effect of K and N on the tree-based ASIF and seek the answer of following questions:

- What is the effect of K and N on a tree-based ASIF?
- What is the combination of K and N that gives the best area-delay product?
- What does the best combination of K and N give when compared to mesh-based ASIF?

In order to determine the effect of LUT and arity size on the tree-based ASIF, experiments are performed for a group of 16 MCNC benchmark circuits and LUT size is varied from 3 to 7 while arity size is varied from 4 to 8 and 16. So with 5 LUT sizes and 6 arity sizes, we have explored a total of 30 architectures where ASIF is generated for each architecture using ASIF-EPER technique as it gives the best overall results compared to other ASIF generation techniques.

Figure 5.11 shows the variation in total area of tree-based ASIF with varying K and N. It can be noted from the figure that for all arity sizes smaller K (e.g. K $=3$ and K $=4$) give better area results compared to bigger ones. Also it can be seen that there is a reduction in total area with increase in N. The results shown in Fig. 5.11 are further elaborated by dividing total area into two parts; logic area and routing area. The effect of varying K and N on the logic area of the architecture is shown in Fig. 5.12. It can be seen from the figure that for all arity sizes logic area increases with increase in LUT size. Figure 5.13 shows that for a fixed N, when K is varied from 3 to 7, there is a gradual decrease in the average number of LUTs that are required by the architecture but there is a sharp rise in the average logic area per LUT as there are 2^k bits in a K-input LUT. So these two effects combined together result in overall increase in logic area of the architecture.

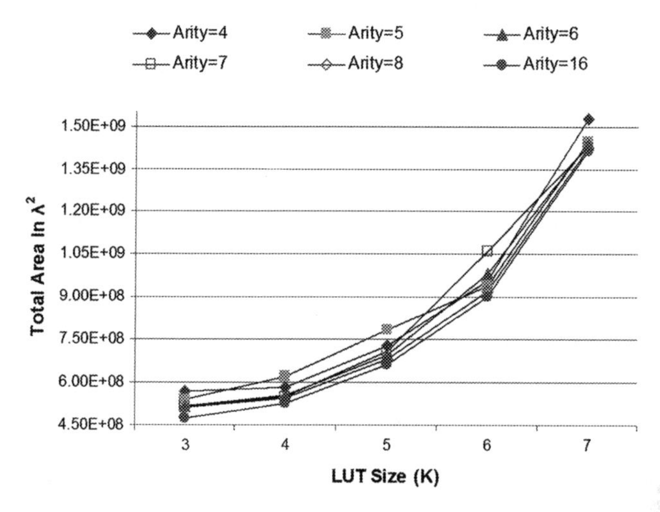

Fig. 5.11 Effect of LUT and arity size on total area of tree-based ASIF

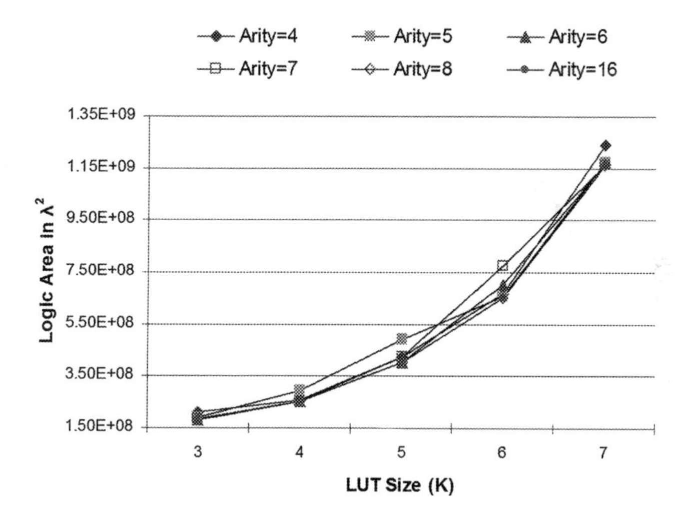

Fig. 5.12 Effect of LUT and arity size on logic area of tree-based ASIF

Figure 5.14 shows variation in routing area with varying LUT and arity sizes. It can be seen from the figure that routing area decreases with increase in K and N. An increase in K decreases the number of LUTs required by the architecture which leads to a smaller architecture and eventually resulting in fewer routing resources required by the architecture. Similarly an increase in N increases the base-cluster size which enables it to absorb more and more signals inside the cluster; hence resulting in reduced routing area. For a fixed N, increase in K reduces average number of LUTs required by the architecture but it increases average routing area per LUT

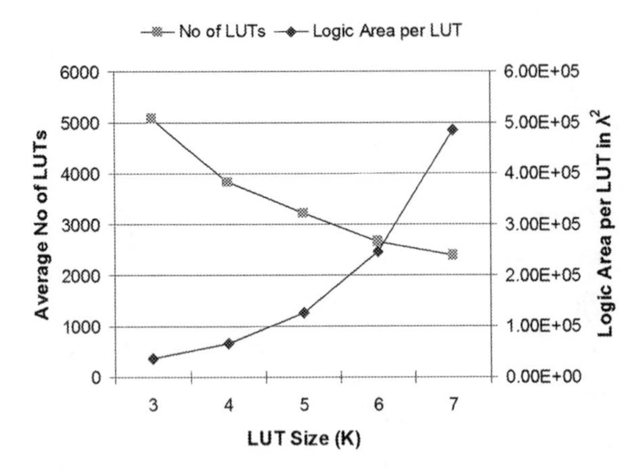

Fig. 5.13 Variation in logic area per LUT for N = 4

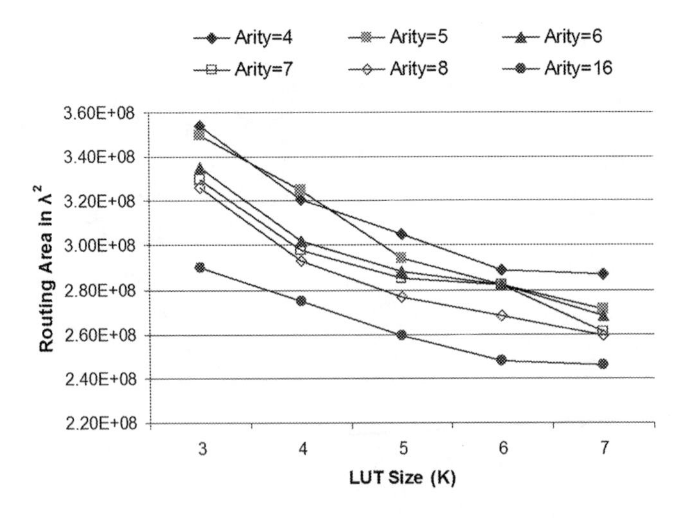

Fig. 5.14 Effect of LUT and arity size on routing area of tree-based ASIF

(refer to Fig. 5.15). However, the decrease in average number of LUTs required by the architecture outweighs the increase in routing area per LUT; hence leading to overall decrease in routing area of the architecture. Although there is a decrease in routing area of the architecture with an increase in LUT and arity size (refer to Fig. 5.14) but this decrease is overshadowed by the increase in the logic area of the architecture (refer to Fig. 5.12); therefore leading to an increased total area of the architecture (refer to Fig. 5.11).

The variation in number of switches crossed by critical path with varying LUT and arity sizes is shown in Fig. 5.16. Number of switches crossed by critical path decrease with an increase in K and N values. K = 7 with N = 16 gives best performance

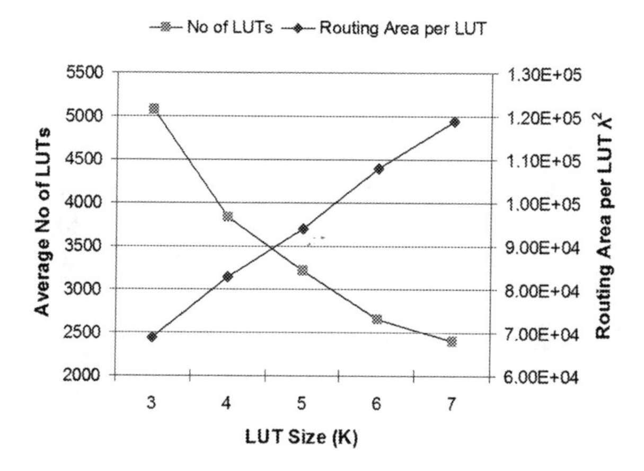

Fig. 5.15 Variation in routing area per LUT for N = 4

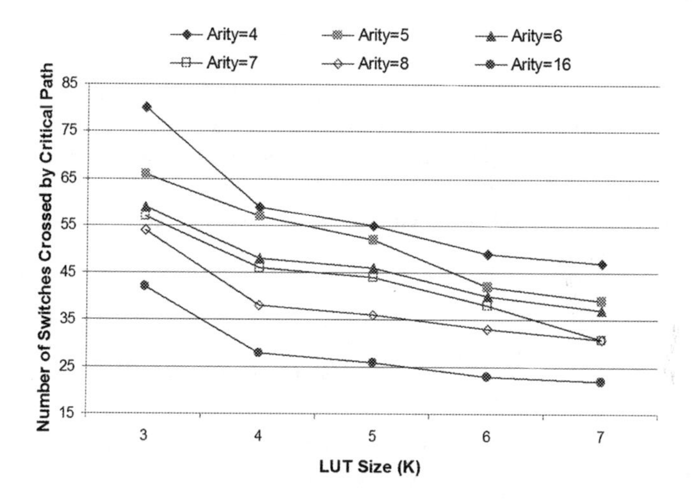

Fig. 5.16 Effect of LUT and arity size on number of switches crossed by critical path

results but the area results of this combination is far worse than the results of smaller LUT sizes. On the other hand, K = 3 with N = 16 produces best area results but its performance results are poor compared to the larger LUT sizes.

So far, we have examined the effect of LUT and arity size on a tree-based ASIF. Results suggest that LUTs with smaller sizes give better area results compared to the LUTs with larger sizes and this trend is reversed in case of performance results. So in order to find the best area-performance results, we have to trade either area with performance or vice-versa. As suggested by [7], we use the area-delay product of different combinations to determine the LUT-arity combination that give best overall results. The area-delay product for different LUT-arity combinations is shown in Fig. 5.17. It can be seen from the figure that for almost all arity sizes, area-delay

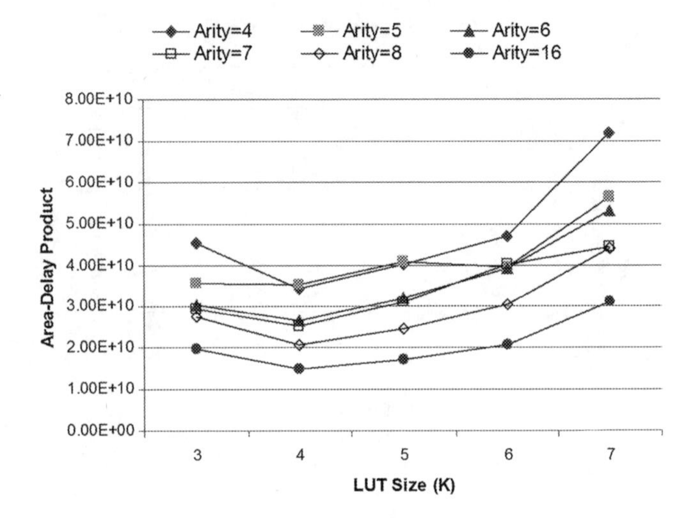

Fig. 5.17 Area delay product for varying LUT and arity size

product initially decreases between K = 3 and K = 4 and then there is a rise in this product. This is mainly because of the sharp rise of area between K = 4 and K = 7 that overshadows the decreased number of switches crossed by critical path. As it can be seen from Fig. 5.17 that K = 4 with N = 16 gives best area-delay product and this is the combination that we will use for the comparison between mesh-based and tree-based ASIFs.

5.6.3 Comparison Between Mesh-Based and Tree-Based ASIFs

Routing area comparison of mesh-based and tree-based ASIF is shown in Fig. 5.18. For tree-based ASIF, LUT size is set to be 4 and arity size is set to be 16 while for equivalent mesh-based ASIF LUT size is set to be 4 and CLB size is 1 for both architectures. Both mesh-based and tree-based ASIFs are generated using ASIF-EPER technique as it gives the best results.

As the logic area of the two architectures is same and it does not change with increase in signal bandwidth, only routing area comparison is presented in Fig. 5.18. Results are presented for a range of signal bandwidths where p values correspond to signal bandwidth for tree-based ASIF and ch-width values correspond to signal bandwidth values for mesh-based ASIF. If we look at the ASIF generation technique of two architectures where all unused routing resources are removed after the placement and routing of netlists, intuition says that the two architecture should have same final area if the logic area of the two architectures is same. But the results in Fig. 5.18 reveal that tree-based ASIF is better than mesh-based ASIF for a range of signal bandwidth values and for maximum signal bandwidth value (i.e. $p = 2.5$ and ch-width = 128) tree-based ASIF gives almost 12% routing area gain compared to mesh-based ASIF.

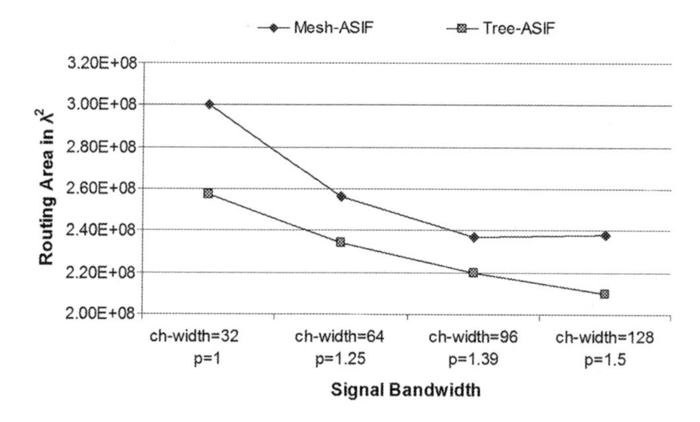

Fig. 5.18 Routing area comparison of mesh-based and tree-based ASIFs

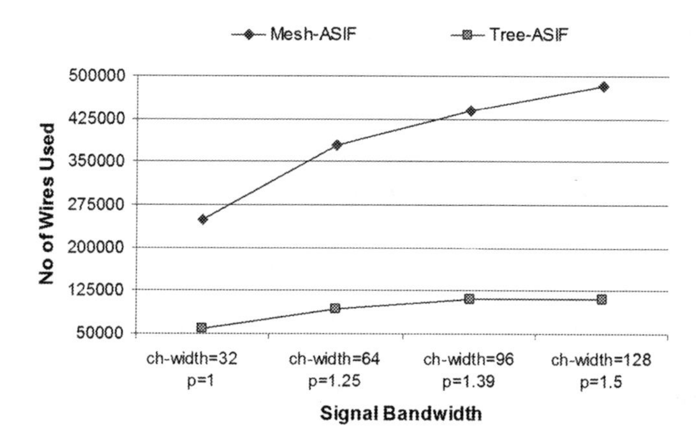

Fig. 5.19 Wire comparison between mesh-based and tree-based ASIFs

This is mainly because of the increased arity size of tree-based ASIF. With increased arity size the partitioner of tree-based ASIF succeeds more easily in performing the inter-netlist optimization without compromising intra-netlist optimization which leads to better partitioning solution and hence more efficient routing resource sharing and reduced area.

Another advantage of tree-based ASIF is that it uses less number of wires than mesh-based ASIF. Wire comparison between mesh-based and tree-based ASIFs for a varying range of signal bandwidth is shown in Fig. 5.19. It can be seen from the figure that for both architectures an increase in signal bandwidth causes an increase in the used number of wires. This is mainly because of efficient routing resource sharing (refer to cases (b) and (d) of Fig. 5.5). However, the increase in wire count for tree-based ASIF is not very sharp and for maximum signal bandwidth it consumes 77% less wires than mesh-based ASIF. The routing wires can play a pivotal role in the area of ASIF if it dominates the logic area. In such a case, mesh-based ASIF might be obliged to use a lower signal bandwidth. However, a tree-based ASIF uses far less wires than mesh-based ASIF and there is a little probability that it may become wire dominant even at higher signal bandwidths. The main reason behind the wire gap is that the mesh-based ASIF is derived from a single length FPGA architecture where no wire crosses more than a single CLB. In case of mesh-based ASIF, these single length wires are converted to long wires if there are no switches involved between them. Since a mesh-based ASIF is generated for a number of predetermined netlists each of which has different routing requirements, it is unlikely to have a large number of long wires in mesh-based ASIF; hence leading to overall increase in total number of wires. On the other hand, generation of tree-based ASIF is based on an architecture that uses a mixture of wires with different lengths. In a tree-based architecture, wire lengths among different cluster of same level are same but they do vary as we move between different levels of hierarchy. So, it would be interesting to generate mesh-based ASIFs that are based on an FPGA architecture having a mixture of wires with varying lengths. However, this exploration is not performed in this work and it is left for future work.

5.6.4 Quality Analysis of Tree-Based ASIF

Since, our ASIF generation techniques are based on heuristic partitioning and routing tools and they do not guarantee the best possible solution. So, the main purpose of this section is to measure and analyze the quality of tools that we use for the ASIF generation.

The quality of a tree-based ASIF can be measured by generating an ASIF for a group of netlists for which an ideal ASIF solution is known. The quality of the generated ASIF can then be measured by comparing it with the ideal solution. Thus, if a tree-based ASIF is generated for similar netlists, ideally speaking, such an ASIF should not require any switch in the routing channel. This ideal ASIF can be achieved if placement of all the netlists is same, and they use exactly the same routing paths to route their connections.

In this work, the quality of ASIF generation methodology is measured by generating ASIF for a group of similar netlists. For this purpose we have chosen "pdc" as our target netlist which is the largest netlist among the 16 MCNC benchmarks shown in Table 5.2. In our analysis, we start with a single netlist and then increment their number by repeatedly placing and routing same netlist to generate the ASIF. The area results for varying number of netlists is shown in Fig. 5.20. The X-axis of the figure shows the number of same netlists used in the generation of ASIF, whereas Y-axis shows the total area of ASIFs. As the number of netlists increase the logic area remains constant, whereas the routing area increases (except for an ideal solution). Figure 5.20 compares five different tree-based ASIFs. These ASIFs are generated by using the best LUT-arity combination and maximum signal bandwidth. Among these ASIFs, ASIF-IPIR (ASIF generated using ideal partitioning and ideal routing) gives

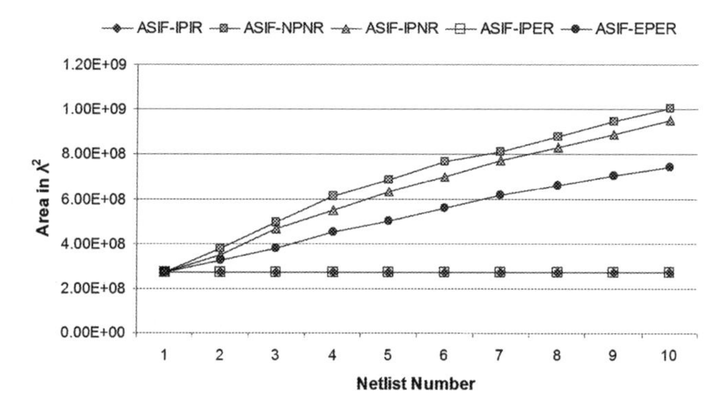

Fig. 5.20 Quality analysis of tree-based ASIF

the best results. This ASIF is generated for the same set of netlists having identical placement and routing. Such an ASIF requires no routing switches. Thus, the area of this ASIF remains constant as the number of netlists increase. This ASIF serves as an ideal solution. Other ASIF generation techniques are compared with this ideal solution. The Ideal placement and routing is achieved manually by using the same placement and routing information for all the netlists.

ASIF-NPNR (ASIF generated using normal partitioning and normal routing) gives the worst results. This ASIF is generated for same set of netlists by using normal partitioning and routing techniques as explained in Sect. 5.4.1. It can be seen in Fig. 5.20 that the ASIF area using this technique increases linearly as the number of netlists increase. ASIF-EPER (ASIF generated using efficient partitioning and efficient routing) gives the results between the best and the worst solution. In this technique, netlists are efficiently partitioned/placed and routed to generate an ASIF-EPER. It is found that an ASIF for 10 "pdc" netlists is almost in the middle of the best and worst solution. ASIF-IPNR (ASIF generated with ideal partitioning and normal routing) is the fourth ASIF that we generate in our quality analysis. As the partitioning is same for all netlists, normal routing will tend to give same routing solution for all netlists. This is because, the congestion driven routing algorithm does not take any random decisions. The congestion of interconnecting nets in a netlist influence the routing solution. Thus, netlists having same partitioning will tend to give same routing solution. To counter this effect, the nets of different netlists are randomly shuffled. Thus, the nets of different netlists are routed in different order, which eventually produces different congestion patterns, and thus different routing solutions. We can see in the figure that an ASIF using ideal partitioning and normal routing is better than the worst solution, but still much far from the ideal solution. ASIF-IPER (ASIF generated using ideal partitioning and efficient routing) is the fifth and final ASIF that we generate for our quality analysis. In this technique the nets of the netlists are randomly shuffled to add random element in routing algorithm. It can be seen in the figure that ideal partitioning with efficient routing produces near ideal results. It affirms that the major loss in ASIF generation is caused by inefficiencies

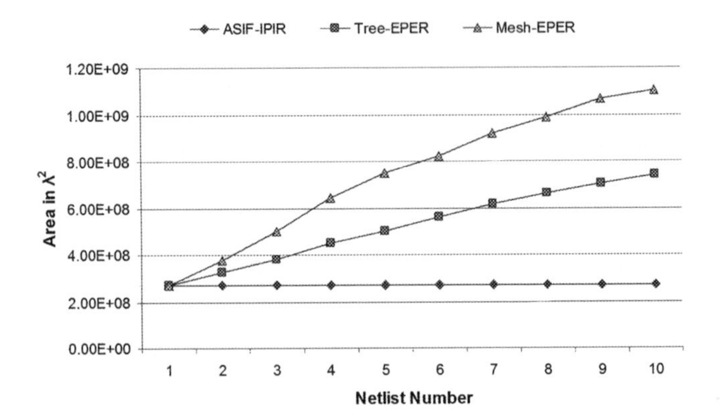

Fig. 5.21 Quality comparison between mesh-based and tree-based ASIFs

in partitioning solution and new techniques are required to be explored to further improve the area of an ASIF.

5.6.5 *Quality Comparison Between Mesh-Based and Tree-Based ASIFs*

In order to evaluate the quality of mesh-based and tree-based architectures, a quality comparison between mesh-based and tree-based ASIFs is performed. For quality comparison, tree-based ASIF uses LUT 4 with arity size 16 and mesh-based ASIF uses LUT 4 whereas the signal bandwidth is set to be maximum that gives best results for both architectures. The quality comparison is performed using ASIF-EPER techniques for mesh-based and tree-based architectures and it is shown in Fig. 5.21. Similar to experiments in previous section, ASIFs are generated for both architectures using "pdc" netlist and results are also compared against an ideal solution. As it can be seen from the figure that for both architectures, the quality of ASIFs deteriorates with increase in the count of netlists. However, tree-based ASIF produces better results as compared to mesh-based ASIF. For small number of netlists, the gap between mesh-based and tree-based ASIFs is not significant but as we increase the count of netlists this gap becomes more and more significant and finally for a group of 10 netlists, tree-based ASIF is around 33% more area efficient than mesh-based ASIF. The gain offered by the tree-based ASIF is because of the structural arrangement of resources of the architecture. In a tree-based architecture, inherently, routing is more predictable and smaller number of paths are available to reach from a source to a destination and this fact compensates the inefficiency of partitioning algorithm. So in a tree-based architecture, every time when a netlist is routed on the architecture, it is more likely that it will follow the same path; hence leading to less number of switches and smaller overall area of the architecture.

5.7 ASIF Hardware Generation

This section details the hardware generation of an ASIF. Similar to the hardware generation of FPGA, the hardware generation of ASIF is integrated with its exploration environment; hence all architectural parameters that are supported by ASIF exploration environment are supported automatically by the ASIF VHDL model generator. The VHDL model of ASIF is passed to Cadence Encounter to generate the layout automatically using 130 nm 6-metal layer CMOS process technology of ST Microelectronics. Main steps involved in the hardware generation of ASIF remain almost same as those involved in FPGA hardware generation. Necessary modifications are however made to meet the architectural needs of ASIF architecture and they are explained below. These modifications are generalized in nature and they are applicable to both mesh-based and tree-based architectures unless otherwise specified.

5.7.1 ASIF Generation Flow

The hardware generation of mesh-based and tree-based ASIFs is based on the architectures described in Sect. 5.2. The generalized flow used for the hardware generation of two architectures is shown in Fig. 5.22. This flow takes architecture description, database of blocks and the netlists as its input. These inputs are then passed to ASIF exploration environment where architecture description and blocks database is used to construct a maximum FPGA architecture that can map any of the given netlists at mutually exclusive times. These netlists are then efficiently placed and routed on the architecture after which all the unused resources of the architecture are removed to generate an ASIF. The software flow generates placement and routing files for each netlist.

An ASIF is represented by the floor-planning of different blocks and the routing graph connecting these blocks. The routing graph of an ASIF is the union of all the routing paths used by any of the mapped netlists. The ASIF VHDL model generator uses the database of different blocks, ASIF floor-planning and ASIF routing graph to generate the ASIF VHDL model.

5.7.2 ASIF VHDL Model Generation

VHDL model of an ASIF is generated in a similar manner as that of FPGA. Similar to an FPGA, the routing network of an ASIF is represented by a routing graph where routing graph is the union of the routing resources that are used by all the netlists mapped on ASIF. The ASIF routing graph contains nodes that are connected through edges; nodes represent a wire, and an edge represent the connections between

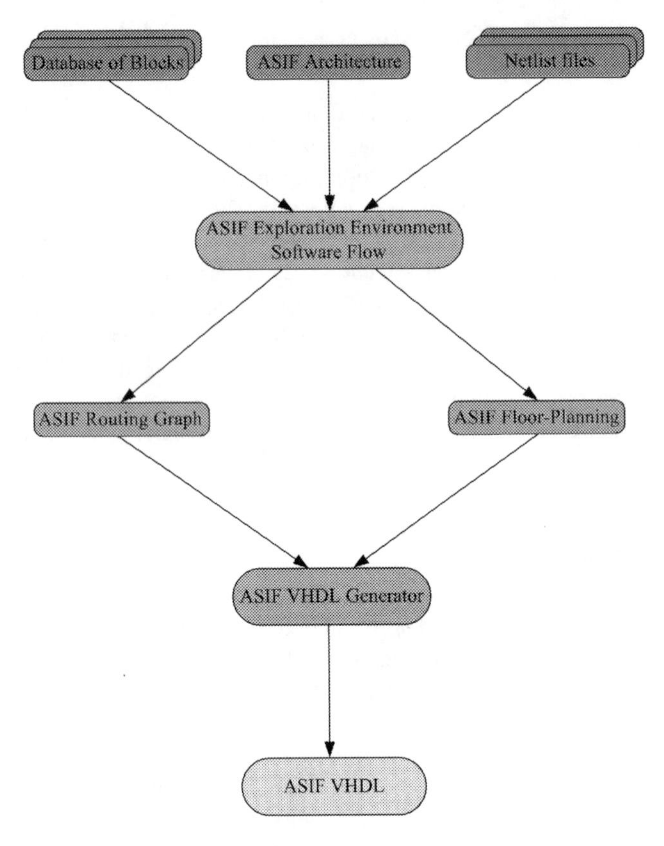

Fig. 5.22 ASIF hardware generation flow

different wires. A wire in the routing graph can be an input or output pin of a block, or a routing wire of the routing network.

An FPGA is transformed to an ASIF by reducing the routing graph of an FPGA. The reduced routing graph of an ASIF is later used to generate the VHDL model. The VHDL generation using routing graph is explained with the help of a small example as shown in Fig. 5.23. Figure 5.23a shows a unidirectional switch box of an FPGA. The routing graph for this switch box is shown in Fig. 5.23b. The physical representation of FPGA switch box is shown in Fig. 5.23c. If a node is driven by more than one nodes, a multiplexor along with the required SRAMs is used to drive multiple nodes to the receiver node. If a node is driven by only a single node, a buffer is used to drive the receiver node. The physical representation of the routing graph is later translated to a VHDL model. This VHDL model is generated using a symbolic standard cell library, SXLIB [9].

Figure 5.23d, g shows routing of two netlists on the FPGA switch box. The wires of the switch box used for routing these netlists are represented as blue lines. The routing graph and physical representation of the switch box, for two routed netlists,

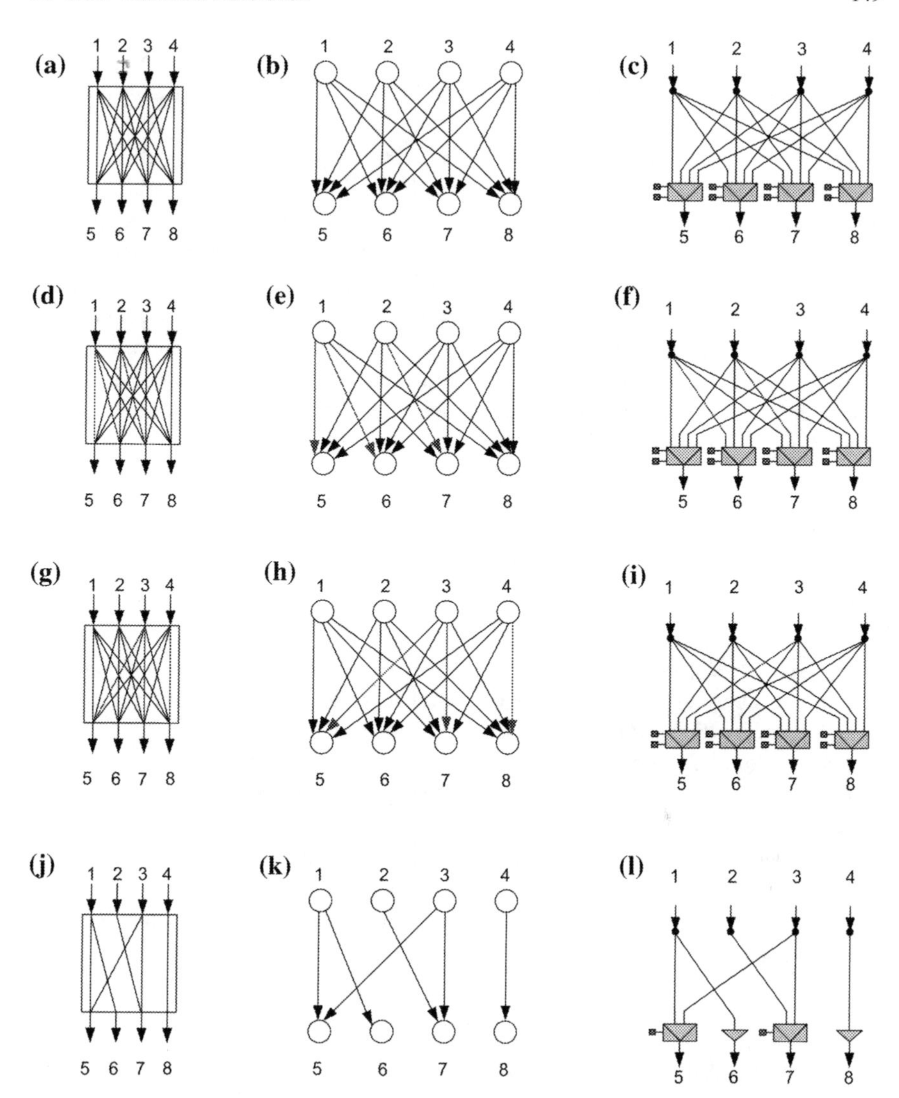

Fig. 5.23 ASIF VHDL model generation part labels. **a** FPGA switch box; **b** routing graph for FPGA switch box; **c** physical representation for FPGA switch box; **d** netlist-1 routed on FPGA switch box; **e** routing graph for netlist-1; **f** physical representation for netlist-1; **g** netlist-2 routed on FGPA switch box; **h** routing graph for netlist-2; **i** physical representation for netlist-2; **j** ASIF switch box for netlist-1, 2; **k** routing graph for ASIF switch box; **l** physical representation for ASIF switch box

are shown in Fig. 5.23e, f, h, i, respectively. An ASIF switch box for the two netlists is shown in Fig. 5.23j. The ASIF switch box only contains those switches and wires that are used by netlist-1 or netlist-2, the remaining switches and wires are removed. The routing graph for ASIF switch box is shown in Fig. 5.23k. It can be seen that the

ASIF routing graph is the sum of the routing paths of all the netlists. The physical representation of ASIF switch box, shown in Fig. 5.23l, is generated by parsing the routing graph of ASIF.

The ASIF routing graph is parsed to generate its physical representation. If a receiver wire is driven by two or more wires, a multiplexor of appropriate size is connected to the receiver wire along with the appropriate SRAM bits. If the receiver wire is being driven by only one wire, a buffer is used to connect the driver wire to the receiver wire. The multiplexors, buffers and SRAMs belong to the same tile to which the receiver wire belongs. The IO and logic block instances are also declared. The input and output pins of these blocks are already represented in the routing graph. Thus, these blocks are automatically linked to physical representation of an ASIF. These blocks are declared in their respective tiles. The VHDL model of these blocks is provided along with the architecture description. The SRAMs used by any logic blocks are also placed in the same way as the SRAMs of routing network are placed.

5.7.3 ASIF Layout Generation

The VHDL model of ASIF is generated using a symbolic standard cell library, SXLIB [9]. The generated VHDL model is translated to 130 nm standard cell library of STMicroelectronics, and then passed to Cadence encounter for layout generation. The ST standard cell library used for experimentation does not contain an SRAM cell, so a LATCH is used. The area of a LATCH is generally larger than that of an SRAM. During the translation to ST cell library, all the buffers are removed. They are later automatically added in the layout phase. The fanout loads are fixed through automatic buffer insertion and gate resizing. The default parameters of encounter enforce at least 5% space reserved for empty space (fillers). ASIF layout generation uses these default parameters.

5.8 Summary and Conclusion

In this chapter, we have presented a new tree-based homogeneous ASIF. An ASIF is a modified form of an FPGA with reduced flexibility and improved density. Four ASIF generation techniques are explored for a set of 16 MCNC benchmarks and comparison of these techniques shows that a tree-based ASIF generated using efficient logic and routing resources sharing technique gives best results. Further the comparison of best ASIF generation techniques with tree-based FPGA shows that the most efficient ASIF generation technique is 64% more area efficient than an equivalent tree-based FPGA. Later, the exploration of LUT and arity size on tree-based ASIF shows that smaller LUTs with larger arity sizes give better area results whereas larger LUTs with larger arity sizes give better performance results. The area-delay product of different LUT-arity combinations reveals that an ASIF with LUT size 4 and arity size 16 produces the most efficient results. Later the comparison between the best techniques of tree-based and mesh-based ASIFs is performed. For tree-based ASIF, LUT 4 arity 16 combination is used while for mesh-based ASIF LUT 4 is used. The two ASIFs are compared over a range of varying signal bandwidth and the results show that the best tree-based ASIF is 12%, 77% more efficient than the best mesh-based ASIF in terms of routing area and number of wires used respectively. Finally the quality comparison of two architectures reveals that for a group of 10 similar netlists, tree-based ASIF is 33% more area efficient than mesh-based ASIF. The work and results presented in this chapter are also published in [117]. In this chapter, we have also presented the model for the automatic hardware generation for mesh-based and tree-based ASIFs. The hardware generator is directly integrated with the exploration environment, thus the VHDL model can be generated for any ASIF architecture supported by the exploration environment.

Chapter 6
Tree-Based ASIF Using Heterogeneous Blocks

An Application Specific Inflexible FPGA (ASIF) is an FPGA with reduced flexibility and improved density that can implement a predetermined set of application circuits which will operate at mutually exclusive times. A homogeneous ASIF is presented in the previous chapter; this chapter extends the work to heterogeneous domain. An ASIF that is reduced from a heterogeneous FPGA is called as heterogeneous ASIF. A heterogeneous ASIF can contain hard-blocks such as multipliers, adders, RAMS etc. In order to generate a heterogeneous ASIF, first a minimal FPGA architecture is defined that can implement any of the applications under consideration. Later these application circuits are efficiently placed and routed to minimize total routing switches required by the heterogeneous FPGA architecture. After that, all unused routing switches are removed from the FPGA to generate a heterogeneous ASIF.

This chapter presents a new tree-based heterogeneous ASIF. Four ASIF generation techniques are explored for tree-based heterogeneous ASIF using 17 benchmarks that use heterogeneous blocks. These benchmarks are further divided into two sets and results show that an ASIF generation technique with efficient logic and routing resource sharing produces optimal results for both benchmark sets. Further experiments are performed to determine the effect of LUT and arity size on tree-based heterogeneous ASIF. Later, tree-based ASIF with the best LUT-arity combination is compared to mesh-based ASIF. Also a quality analysis of tree-based heterogeneous ASIF and a quality comparison between mesh-based and tree-based ASIFs is performed.

6.1 Reference Heterogeneous FPGA Architectures

This section gives a brief overview of heterogeneous mesh-based and tree-based FPGA architectures and the related software flow that is used to place and route different benchmarks on these architectures. Application circuits are efficiently placed and routed on these architectures and later they are reduced to their respective ASIFs.

U. Farooq et al., *Tree-Based Heterogeneous FPGA Architectures*,
DOI: 10.1007/978-1-4614-3594-5_6,

6.1.1 Heterogeneous Tree-Based FPGA Architecture

A tree-based heterogeneous architecture [45] is a hierarchical architecture having unidirectional interconnect. In this architecture CLBs, I/Os and HBs are partitioned into a multilevel clustered structure where each cluster contains sub-clusters and switch blocks allow to connect external signals to sub-clusters. The number of signals entering into and leaving from the cluster can be varied depending upon the netlist requirement. The signal bandwidth of clusters is controlled using Rent's rule [74] which is easily adapted to tree-based architecture. This rule states that

$$IO = \left(\underbrace{k.n^{\ell}}_{L.B(p)} + \underbrace{\sum_{x=1}^{z} a_x.b_x.n^{(\ell-\ell_x)}}_{H.B(p)} \right)^{p} \tag{6.1}$$

where

$$H.B(p) = \begin{Bmatrix} 0 & if(\ell - \ell_x < 0) \\ a_x.b_x.n^{(\ell-\ell_x)} & if(\ell - \ell_x \geq 0) \end{Bmatrix} \tag{6.2}$$

In Eq. 6.1 ℓ is a tree level, n is the arity size, k is the number of in/out pins of a LUT, a_x is the number of in/out pins of a HB, ℓ_x is the level where HB is located, b_x is the number of HBs at the level where it is located, z is the number of types of HBs supported by the architecture and IO is the number of in/out pins of a cluster at level ℓ. Since there can be more than one type of HBs, their contribution is accumulated and then added to the $L.B(p)$ part of Eq. 6.1 to calculate p. The value of p is a factor that determines the cluster bandwidth at each level of the tree-based architecture and it is averaged across all the levels to determine the p for the architecture. Further details regarding the tree-based heterogeneous FPGA architecture can be found in Chap. 4.

6.1.2 Heterogeneous Mesh-Based FPGA Architecture

The architecture used in this work is a VPR-style (Versatile Place & Route) [81] architecture that contains configurable logic blocks (CLBs), I/Os and hard-blocks (HBs) that are arranged on a two dimensional grid. In order to incorporate HBs in a mesh-based FPGA, the size of HBs is quantized with size of the smallest block of the architecture i.e. CLB. The width and height of an HB is therefore a multiple of width and height of the smallest block in the architecture. Further details regarding the architecture can be found in Sect. 4.2.2.

6.1.3 Software Flow

The software flow used to place and route different benchmarks (netlists) on the two architectures is same as discussed in Chap. 4. The flow starts with the conversion of vst (structured vhdl) file to BLIF format [20]. The BLIF file is then passed through a parser which removes HBs from the file and passes it to SIS [14] that synthesizes it into LUT format of a given size (K). The file is then passed to T-VPACK [14] which packs and converts it into net format. A netlist in net format contains CLBs and I/O instances that are connected together using nets. The size of a CLB is defined as the number of LUTs contained in it and in this work this size is set to be 1 for both mesh-based and tree-based architectures. After T-VPACK the netlist is passed through another parser that adds previously removed HBs and finally it is placed and routed separately on tree-based and mesh-based FPGAs. A detailed description regarding the placement and routing of netlists on the two architectures is already given in Sect. 4.4.

6.2 Heterogeneous ASIF Generation Techniques

A heterogeneous ASIF is a modified form of a heterogeneous FPGA architecture. A heterogeneous ASIF has reduced flexibility but improved density compared to heterogeneous FPGA and it can contain hard-blocks like multiplier, adders, RAMs etc. A group of netlists are placed and routed on the FPGA architecture. Different automatic floor-planning techniques are used to optimize the position of different blocks on the architecture. Among these floor-planning techniques, the technique Block-Move-Rotate (BMR) gives the best results (refer to Chap. 4) and this is the technique that is used for the generation of mesh-based heterogeneous ASIF. Similarly, for tree-based ASIF ASYM techniques gives best results (refer to Chap. 4) and this is the technique that is used for tree-based ASIF.

In order to generate a mesh-based heterogeneous ASIF, initially a heterogeneous FPGA architecture is defined using an architecture description file. Blocks of different sizes are defined, and later mapped on a grid of equally sized slots. The placer maps multiple netlists together to get a single architecture floor-planning for all netlists. Efficient placement and routing techniques are exactly the same as discussed in previous chapter. However, necessary changes are performed to handle heterogeneous blocks. Efficient placement tries to place the instances of different netlists in such a way that minimum routing switches are required in an FPGA. Later, efficient routing encourages different netlists to route their nets on an FPGA with maximum common routing paths. After all netlists are efficiently placed and routed on FPGA, unused switches are removed from the architecture to generate a heterogeneous ASIF.

For tree-based ASIF, similar to mesh-based architecture, initially an FPGA architecture is defined using an architecture description file. Architecture description file contains complete detail about different parameters of the architecture including the

Table 6.1 SET I benchmark details

Index	Circuit name	In	Out	Mult 18×18	LUT-3	LUT-7
1.	cf_fir_3_8_8_ut	42	22	4	217	186
2.	diffeq_f_systemC	66	99	4	2,114	1,366
3.	diffeq_paj_convert	12	101	5	1,013	471
4.	fir_scu	10	27	17	2,267	629
5.	iir	33	30	5	524	322
6.	iir1	28	15	5	898	410
7.	rs_decoder_1	13	20	13	2,367	900
8.	rs_decoder_2	21	20	9	4,204	1,835
9.	Maximum	66	101	17	4,204	1,835

signal bandwidth of different clusters located at different levels of hierarchy, types of different blocks, the level where they are located and their pin details etc. Once the architecture is defined, netlists are efficiently partitioned and then placed and routed on the architecture. Different ASIF generation techniques are employed for tree-based architecture in a manner similar to the previous chapter except that necessary modification are made to handle the heterogeneity of the architecture. Once the netlists are efficiently placed and routed on the architecture, unused resources of the architecture are removed to generate an ASIF.

6.3 Experimentation and Analysis

6.3.1 Experimental Benchmarks

This work uses 17 open core benchmarks to explore the effect of different ASIF generation techniques on the tree-based ASIF. These benchmarks are further divided into two sets. The division of benchmarks into two sets is basically based on the type of hard-blocks that are supported by these benchmarks. The details of two sets of benchmarks is shown in Tables 6.1 and 6.2 respectively. As it can be seen from these tables that one set supports only 18×18 multipliers while the second set supports 16 × 16 multipliers along with 20 + 20 adders. Apart from hard-blocks, these benchmarks also contain logic blocks; size of whom in terms of different LUT sizes is shown in the last two columns of these tables. These benchmarks are converted into .net format using the same flow as described in Chap. 4. Netlists with different LUT sizes are generated where number of I/Os and hard-blocks remain unchanged, but the number of LUTs change with change in LUT size. Details of netlists for two sets with LUT size 3 and 7 are shown in last two columns of Tables 6.1 and 6.2 respectively.

Table 6.2 SET II benchmark details

Index	Circuit name	In	Out	Mult 16×16	Add 20+20	LUT-3	LUT-7
1.	cf_fir_3_8_8_open	42	18	4	3	159	159
2.	cf_fir_7_16_16	146	35	8	14	639	638
3.	cfft16x8	20	40	–	26	1,937	1,122
4.	cordic_p2r	18	32	–	43	803	801
5.	cordic_r2p	34	40	–	52	1,497	1,178
6.	fm	9	12	1	19	1,508	1,046
7.	fm_receiver	10	12	1	20	1,004	677
8.	lms	18	16	10	11	965	935
9.	reed_solomon	138	128	16	16	537	537
10.	Maximum	146	128	16	52	1,937	1,178

This section is basically divided into three parts. First the effect of different ASIF generation techniques on tree-based heterogeneous ASIF is explored for both sets of benchmarks. For each ASIF generation technique, separate ASIFs are generated for both sets of benchmarks. Results of different ASIF generation techniques are then compared and later results of ASIF generation techniques are also compared with those of equivalent tree-based heterogeneous FPGA. Secondly, experiments are performed to determine the effect of LUT and arity size on tree-based heterogeneous ASIF. Experiments are performed for a range of architectures that are generated using different combinations of LUT and arity sizes and netlists are placed and routed on them. Finally, after determining the best LUT and arity size combination for tree-based ASIF, a comparison between mesh-based and tree-based ASIFs is performed using their best ASIF generation techniques.

6.3.2 Effect of Different ASIF Generation Techniques on Heterogeneous Tree-Based ASIF

Figure 6.1 shows the variation in routing area of tree-based architecture using different ASIF generation techniques. In this experimentation, for all ASIF generation techniques, LUT size is set to be 4, arity size is 4, and CLB size is 1. In this figure columns 1–5 correspond to results for SET I benchmarks whereas columns 6 to 10 correspond to results for SET II benchmarks. For each set of benchmarks, ASIFs of four different kinds are generated over a range of signal bandwidths. The range of signal bandwidth is varied from the minimum required to the maximum value beyond which a further increase in signal bandwidth does not improve the routing area of ASIF. It can be seen from the figure that for each ASIF generation technique (for both sets), an increase in signal bandwidth decreases the routing area of architecture. Although the rate of decrease in routing area of the architecture with increasing signal bandwidth varies, a decrease in the routing area of the architecture for each ASIF generation technique remains valid. The difference in the routing areas

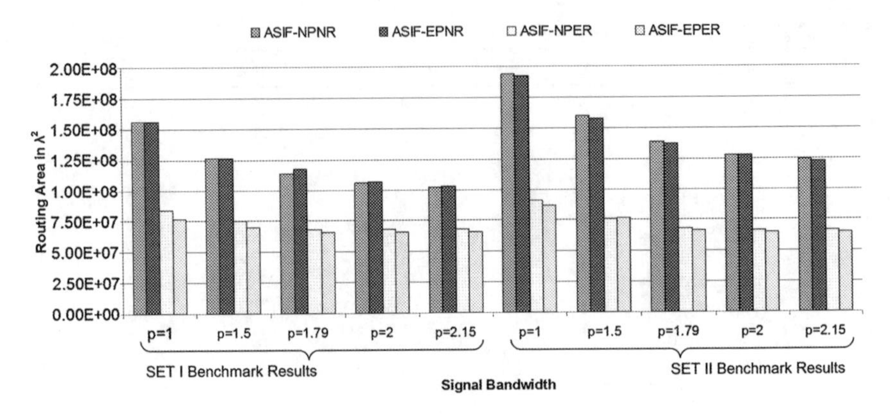

Fig. 6.1 Routing area comparison between different tree-based heterogeneous ASIFs

of different ASIF generation techniques is mainly because of the different optimization techniques that are employed in each ASIF generation technique. As can be seen from the figure that the ASIF generation technique with no optimization (i.e. ASIF-NPNR) produces the worst results among the four ASIF generation techniques for both sets. Also the technique with only efficient partitioning solution (i.e. ASIF-EPNR) is not as efficient as the technique with only efficient routing solution (i.e. ASIF-NPER). Efficient partitioning alone is unable to produce any significant results because it is unable to exploit the routing resources of the architecture that contribute a major percentage of the total area. However, when it is combined with efficient routing, it gives the best overall results compared to the rest of ASIF generation techniques.

Figure 6.2 shows the variation in the number of wires that are used by different ASIF generation techniques. Similar to area results, first 5 columns give results for SET I while the next 5 columns give results for SET II benchmarks. For both sets, signal bandwidth is varied from the minimum required to the maximum, and results show that, at maximum signal bandwidth, ASIF-EPNR uses the least number of wires while ASIF-NPER uses the most number of wires for both sets. Similar to the results for homogeneous ASIFs, the techniques using efficient routing resource sharing consume more wires as compared to the techniques that are not using efficient routing resource sharing. Similarly, techniques with efficient logic resource sharing consume less number of wires than the techniques that are not using efficient logic sharing. These two facts combined together result in the trends that are shown in Fig. 6.2.

Figure 6.3 shows the comparison between different ASIF generation techniques for different number of netlists. In this figure, first 8 columns give results for SET I and next 9 columns give results for SET II respectively. For fist 8 columns, column 1 means that ASIF is generated for first netlist of Table 6.1, column 2 means that ASIF is generated for first two netlists of the table and same rule applies to the netlists of Table 6.2. For all four techniques, the signal bandwidth is set to be maximum

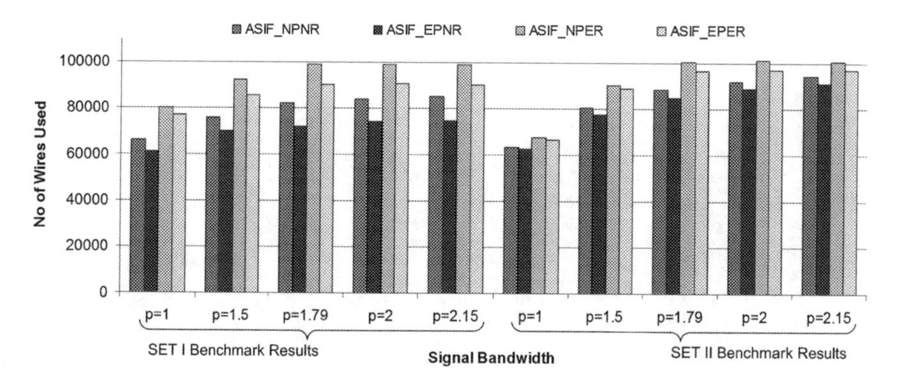

Fig. 6.2 Wire comparison between different heterogeneous ASIF techniques

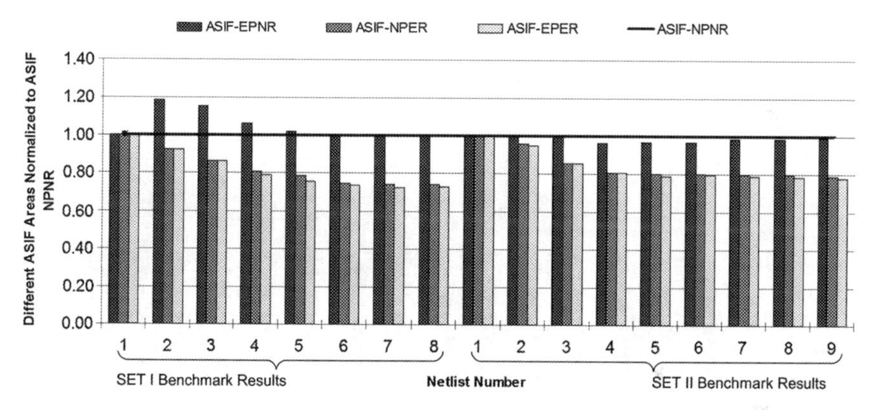

Fig. 6.3 Different ASIFs normalized to ASIF-NPNR

that produces the best results. As it can be seen from the figure that ASIF-NPNR gives worst results and ASIF-EPER gives best results in terms of area. Further the gap between the best and the worst ASIF techniques increases with increase in the number of netlists for which the ASIF is being generated.

In order to perform the performance comparison of different ASIF generation techniques, we have employed a model that calculates the number of switches that are crossed by different applications mapped on ASIF. Figure 6.4 shows the performance comparison results between different techniques for a varying range of signal bandwidths. In this figure results of first 5 columns correspond to SET I benchmarks while remaining 5 correspond to SET II benchmarks. Since each ASIF is generated for a group of netlists and each netlist crosses different number of switches on its critical path, an average of critical path switch number is presented here for each technique. It can be seen from the figure that for both sets of benchmarks and for all generation techniques, an increase in signal bandwidth results in a decrease in average number of switches that are crossed by critical path. These results are in direct correspondence with those of Fig. 6.1 and further this observation is enhanced

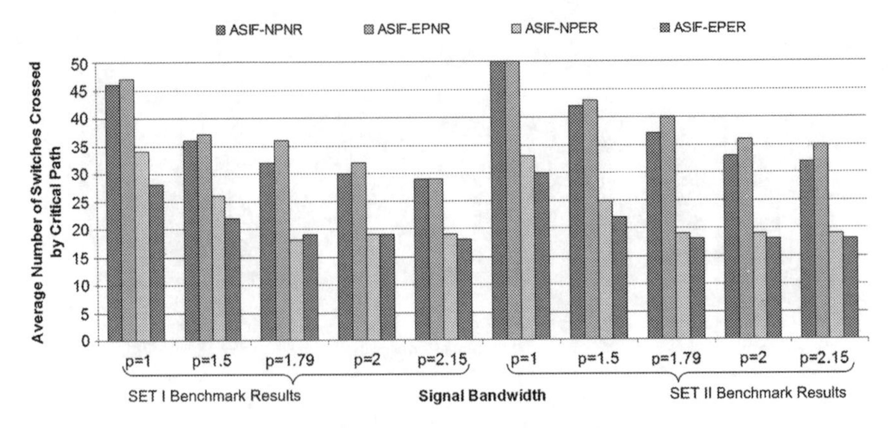

Fig. 6.4 Performance comparison between different tree-based heterogeneous ASIFs

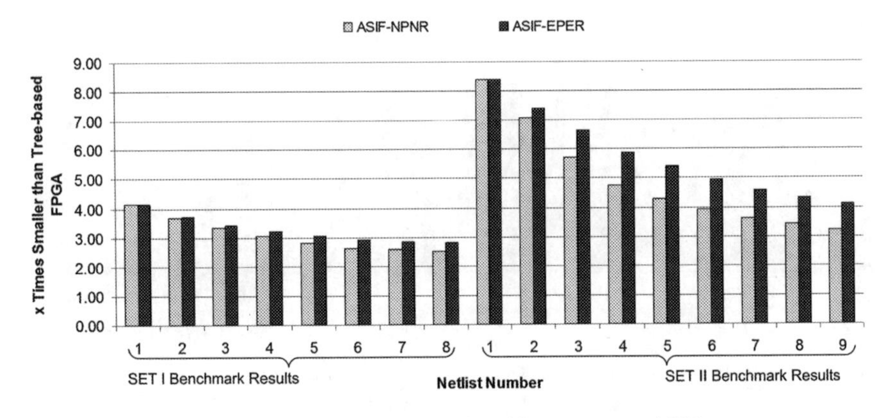

Fig. 6.5 Tree-based heterogeneous FPGA vs tree-based heterogeneous ASIFs

by the fact that ASIF-EPER produces the best results in both cases. Although the performance comparison presented here is based on a simple model and it does not give the exact temporal delay, yet it gives us an idea about the performance of each ASIF generation technique.

Figure 6.5 shows the area comparison between tree-based FPGA and the worst and the best ASIF generation techniques. In this figure first 8 columns give results for SET I benchmarks and remaining 9 columns give results for SET II benchmarks respectively. Signal bandwidth is set to be maximum for ASIF while equivalent FPGA has minimum signal bandwidth which is capable of mapping any of the benchmarks of respective sets. In this figure, starting from 1, ASIFs are generated for a varying range of netlists and results are then compared to those of tree-based FPGA. For SET I benchmarks, when varying the number of netlists, the order shown in Table 6.1 is followed. For example, netlist 1 means an ASIF is generated for "cf_fir_3_8_8_ut" only and netlist 2 means that an ASIF is generated for "cf_fir_3_8_8_ut" and "diffeq_f_SystemC" and so on. Similarly, for SET II benchmarks, the order shown in Table 6.2 is followed. Results of ASIFs of both sets for varying number of netlists are compared against their respective LUT-4, arity-4 tree-based FPGAs that can implement any netlist of the Table 6.1 for SET I and any netlist of Table 6.2 for SET II respectively. It can be seen from the figure that for best ASIF generation technique, for SET I, for 1 netlist ASIF is 4.14 times or 76% smaller than tree-based FPGA. However as the number of netlists increase, the gap between the two decreases and for 8 netlists ASIF is 2.8 times or 64% smaller than the tree-based FPGA. Similarly for SET II benchmarks, for 1 netlist the best ASIF is 8.4 times smaller than the corresponding tree-based FPGA and when the count of netlists is increased to 9 the tree-based ASIF is 4.12 times or 75% smaller than the corresponding tree-based FPGA. The results in Fig. 6.5 suggest that ASIF gives excellent area gain for a small group of netlist. However, the advantage of ASIF begins to fade as the size of the set of netlists increases and for a very large group of netlists it might not be able to give any significant area advantages.

So far in this section we have seen that ASIF-EPER produces best results for both sets of benchmarks that are under consideration. Although we have fixed LUT and arity sizes to 4, ASIF-EPER gives the best results even if we change this combination; hence this is the technique that we will use to determine the effect of LUT and arity size on a tree-based heterogeneous ASIF.

6.3.3 Effect of LUT and Arity Size on Heterogeneous Tree-Based ASIF

In order to determine the effect of LUT and arity size on a tree-based heterogeneous ASIF, experiments are performed for a group of 17 open core benchmarks and these benchmarks are further divided into two sets (refer to Tables 6.1 and 6.2). For benchmarks of SET I and SET II, LUT size is varied from 3 to 7 while arity size is varied from 4 to 8 and 16. So with 5 LUT sizes and 6 arity sizes, we have explored 30 architectures for SET I and SET II where ASIF is generated for each architecture using the ASIF-EPER technique as it gives the best overall results compared to other ASIF generation techniques.

Effect of varying LUT and arity size on total area of tree-based heterogeneous ASIF is shown in Fig. 6.6 where first 5 columns give results for SET I and next 5 columns give results for SET II. It can be seen from the figure that smaller LUTs give better area results as compared to the larger LUTs. Also it can be noticed from the figure that, in general, for a fixed LUT size the total area of the ASIF decreases with an increase in the arity size. On the other hand, for a fixed arity size, an increase in LUT size increases the total area of the architecture. The observations made in this figure are further elaborated using Figs. 6.7 and 6.8.

Figure 6.7 shows the effect of varying K and N on the logic area of tree-based heterogeneous ASIF. It can be seen from the figure that for all arity sizes, an increase

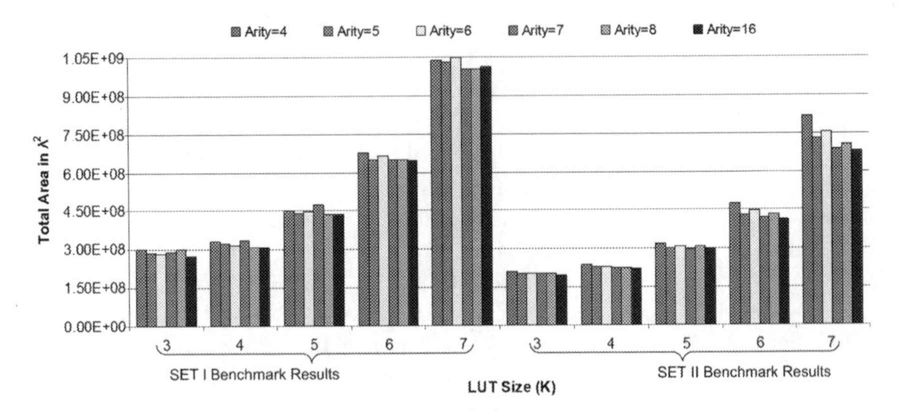

Fig. 6.6 Effect of LUT and arity size on total area of tree-based heterogeneous ASIF

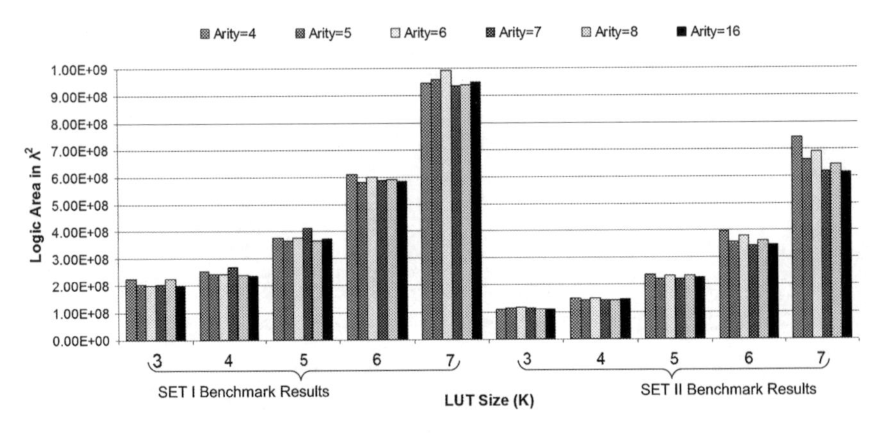

Fig. 6.7 Effect of LUT and arity size on logic area of tree-based heterogeneous ASIF

in K increases the logic area of the architecture. In fact, for a fixed arity size, increase in LUT size decreases the total number of LUTs required. But at the same time LUT area increases exponentially with increase in its size. Normally the increase in LUT area overshadows the decrease in its number; hence causing a sharp rise in the logic area of the architecture.

Figure 6.8 shows the effect of varying K and N on the routing area of tree-based heterogeneous ASIF. It can be seen from the figure that for all arity sizes, in general, a raise in K causes a decline in the routing area of the architecture. For a fixed N, elevation in K reduces average number of LUTs required by the architecture but it raises average routing area per LUT. However, the fall in average number of LUTs required by the architecture outweighs the rise in routing area per LUT; hence leading to overall decline in routing area of the architecture. Although there is a decrease in routing area of the architecture (refer to Fig. 6.8) with an increase in LUT and arity size but this decline is overshadowed by the increase in the logic area of the

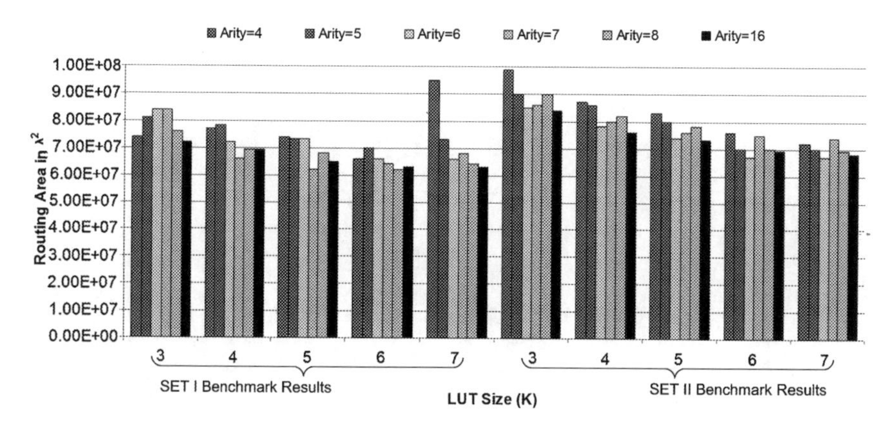

Fig. 6.8 Effect of LUT and arity size on routing area of tree-based heterogeneous ASIF

architecture (refer to Fig. 6.7); therefore leading to an increased total area of the architecture (refer to Fig. 6.6).

Figure 6.9 shows the effect of LUT and arity size on the critical path switch number of a tree-based heterogeneous ASIF. This work considers only an elementary model to have an idea about the critical path delay and wire delays are not considered here. In Fig. 6.9 columns 1–5 give the results for SET I benchmarks while columns 6–10 give results for SET II benchmarks. It can be seen from the figure that an increase in LUT and arity size decreases the number of switches that are crossed by critical path. For both sets of benchmarks, for a fixed K, an increase in N decreases the average number of switches crossed by critical path. Similarly, for a fixed N, an increase in K decreases the average number of switches crossed by critical path. However, for a fixed N, the two benchmark sets have different rates of decrease in average critical path switch number with increasing K. For example in case of SET I, for N=4, as the K is varied from 3 to 7, a 28% decline in average switch number is observed. On the other hand, in case of SET II, a decline of only 7% is observed for N=4 and variation in K from 3 to 7. This is because of the fact that for a fixed N, when K is varied from 3 to 7, for 5 out of 9 benchmarks of SET II, LUT requirement remains almost same (refer to Table 6.2). Number of switches of these 5 benchmarks constitute around 88% of the average switch numbers of SET II; hence a significant decrease in the average switch numbers is not observed. Figures 6.6 to 6.9 give an overview about the effect of K and N on tree-based ASIF. Results shown in these figures imply that ASIFs with small K and large N (K = 3 or 4 and N = 16) give best area results and large K with large N give best performance results. So, we have to find a trade-off between area and performance results. For this purpose, area-delay products of ASIFs with varying LUT and arity size are shown in Fig. 6.10. In this figure, first five columns give results for SET I benchmarks while next five columns give results for SET II benchmarks. It can be seen from the figure that K = 4 with N = 16 gives best results for SET I benchmarks and K = 3 with N = 16 gives best results for SET II benchmarks. Since, we have determined the best LUT-arity

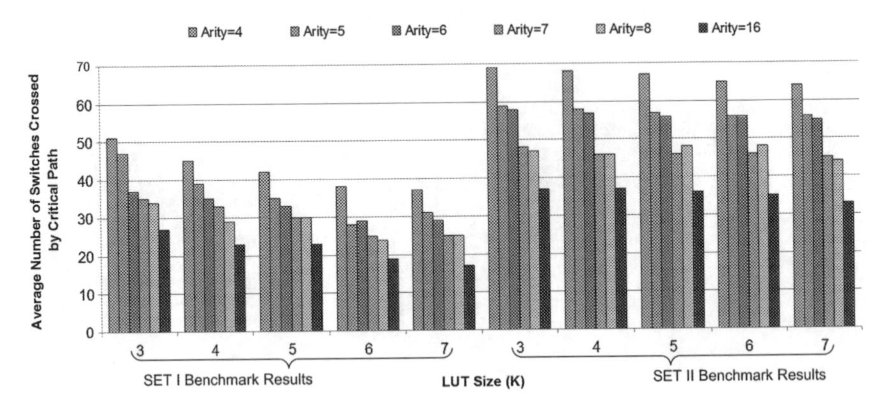

Fig. 6.9 Effect of LUT and arity size on critical path of tree-based heterogeneous ASIF

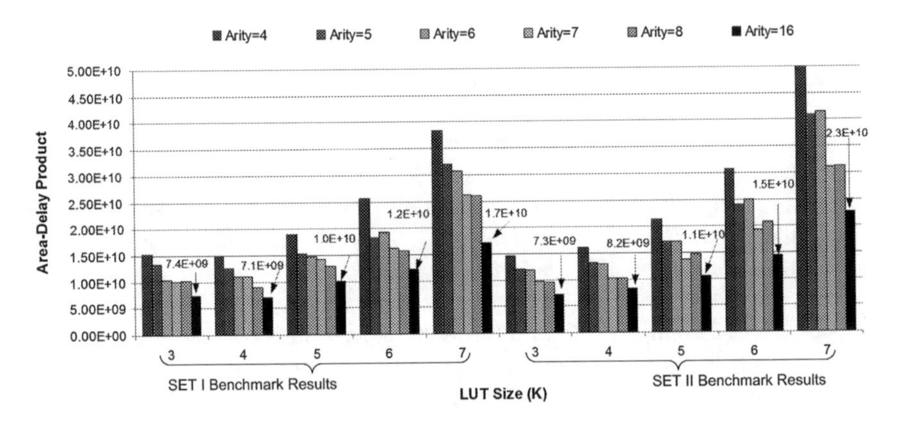

Fig. 6.10 Effect of LUT and arity size on area-delay product of tree-based heterogeneous ASIF

combinations for the two sets, these combinations will be used for comparison of heterogeneous mesh-based and tree-based ASIFs.

6.3.4 Comparison Between Heterogeneous Mesh-Based and Tree-Based ASIFs

Now that we have determined the best LUT-arity combination for tree-based ASIF, here we present a comparison between the best techniques of tree-based ASIF and equivalent mesh-based ASIF. The comparison results of heterogeneous mesh-based and tree-based ASIFs are shown in Fig. 6.11. In this figure columns 1–4 present routing area comparison between mesh and tree architectures for SET I benchmarks whereas columns 5–8 present comparison results for SET II benchmarks. For both

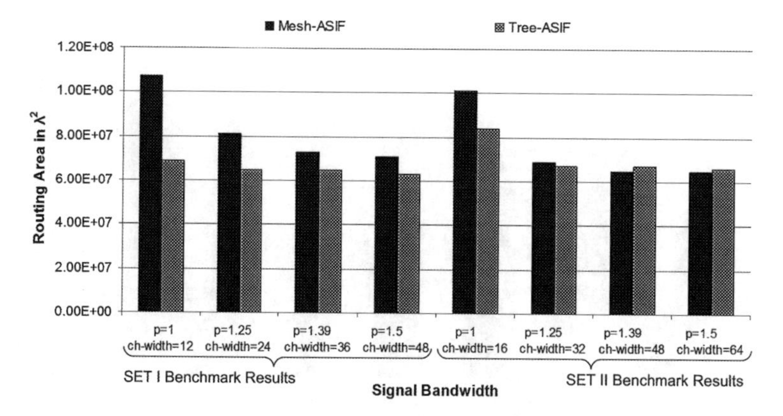

Fig. 6.11 Routing area comparison between heterogeneous mesh-based and tree-based ASIFs

sets, mesh-based and tree-based ASIFs are generated using ASIF-EPER technique as it gives the best results. Signal bandwidth is varied for both sets starting from a minimum required to a maximum value. An increase of signal bandwidth beyond maximum value does not improve further the routing area of the architecture. For tree-based ASIF, for SET I, LUT size is set to be 4 and arity size is set to be 16 and for SET II LUT size is set to be 3 and arity size is set to be 16. In case of mesh-based ASIF, for SET I, LUT size is 4 and for SET II LUT size is 3 while CLB size is 1 for both architectures.

Since the variation in signal bandwidth does not affect the logic area, only routing area results are presented in Fig. 6.11. In this figure "p" values correspond to the signal bandwidth of tree-based ASIF and it is calculated using Eq. 6.1 whereas "ch-width" values correspond to the signal bandwidth values of mesh-based ASIF. It can be seen from the figure that for SET I, tree-based ASIF produces better results when compared to mesh-based ASIF for all signal bandwidths. For maximum signal bandwidth (i.e. $p = 1.5$, ch-width $= 48$) tree-based ASIF gives 11.27% better routing area than mesh-based ASIF. For SET II benchmarks, however, tree-based ASIF gives better routing area results for smaller signal bandwidths. But with larger bandwidth tree-based ASIF looses the gain and for a maximum signal bandwidth value mesh-based ASIF is 1.5% better than tree-based ASIF. Intuitively if we look at the ASIF generation mechanism of the two architectures, they should give same results as unused resources are removed in both architectures after the mapping of benchmarks under consideration. However, it is the arrangement of resources and quality of tools that makes the difference. For example for mesh-based architecture, the ASIF generation mechanism uses BMR floor-planning at its core (for details regarding BMR floor-planning refer to Chap. 4). This is the most flexible floor-planning technique where hard-blocks are allowed to move and rotate and this is the reason that mesh-based ASIF performs better in case of SET II benchmarks. Since, for SET II benchmarks, the major percentage of total communication is between CLBs and HBs and BMR is flexible enough to take full advantage of this trend. However, for

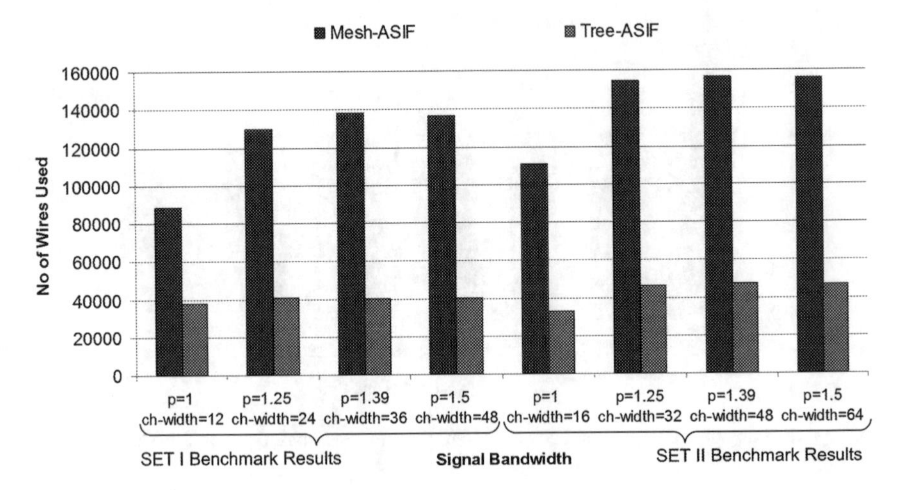

Fig. 6.12 Wire comparison between heterogeneous mesh-based and tree-based ASIFs

tree-based ASIF, ASYM exploration technique is used (for details regarding ASYM technique refer to Chap. 4) which is comparatively less flexible than BMR; hence causing an increased area than mesh-based ASIF. For SET I benchmarks however, the major percentage of total communication is between the CLBs and tree-based ASIF is not affected in that case.

Area results are not the only comparison performed between mesh-based and tree-based ASIFs, the two architectures are also compared on the basis of number of wires consumed by the two architectures and the results are shown in Fig. 6.12. Similar to area results, first four columns present comparison results for SET I benchmarks while last four present results for SET II benchmarks. As it can be seen from the figure that for both sets of benchmarks, tree-based ASIF consumes far less wires than mesh-based ASIF. For maximum signal bandwidth values, tree-based ASIF uses 70, 69% less wires than mesh-based ASIF. The number of wires can play a pivotal role while making a choice for the architecture. For mesh-based ASIF, the number of wires are high and they increase significantly with the increase in the signal bandwidth and at higher bandwidth values the architecture might become wire dominant and it might be obliged to use a lower signal bandwidth. So in such a case, tree-based ASIF is clearly a better choice when compared to mesh-based ASIF as it gives relatively better area and consumes significantly less wires than mesh-based ASIF.

6.4 Quality Analysis of Heterogeneous Tree-Based ASIF

Similar to the quality analysis of tree-based homogeneous ASIF, the quality analysis of tree-based heterogeneous ASIF is also performed. Quality analysis of tree-based heterogeneous ASIF is performed for two sets separately and results for SET I and

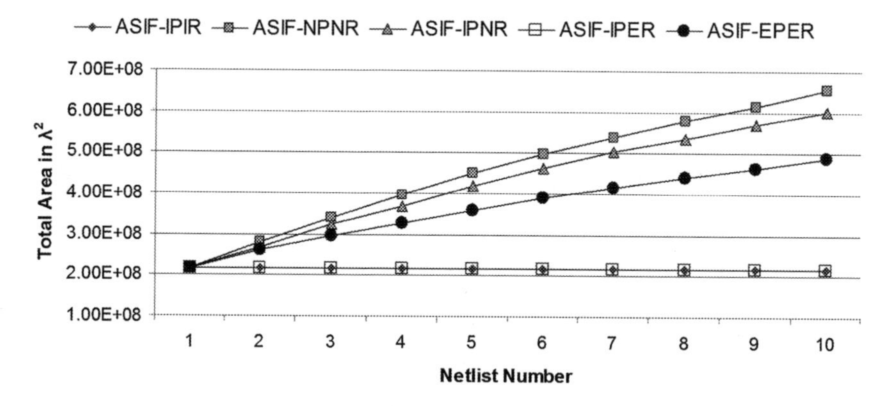

Fig. 6.13 Quality analysis of tree-based heterogeneous ASIF (SET I)

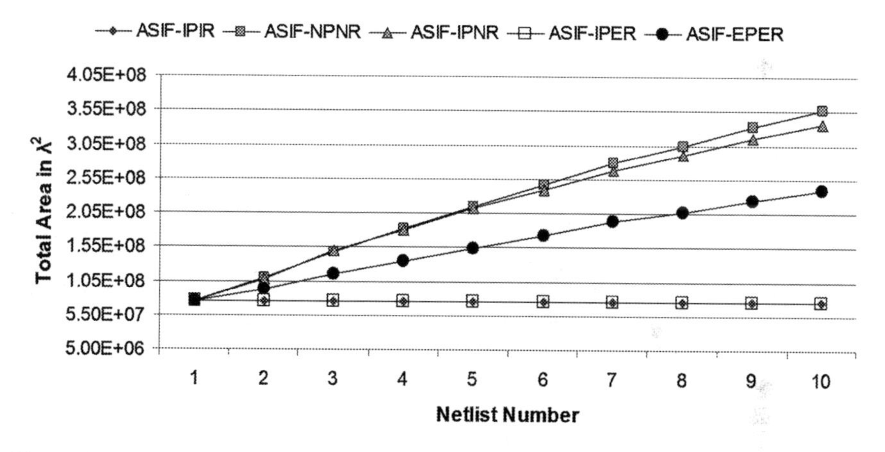

Fig. 6.14 Quality analysis of tree-based heterogeneous ASIF (SET II)

SET II are shown in Figs. 6.13 and 6.14 respectively. In these figures, the X-axis shows the number of same netlists used in the generation of ASIF, whereas Y-axis shows the total area of ASIFs. For SET I, quality analysis is performed using "rs_decoder_2" and for SET II quality analysis is performed using "cfft16x8". The two netlists are chosen as they are the biggest among the two sets. Similar to the quality analysis of homogeneous ASIF, five different techniques are used to perform the quality analysis of tree-based heterogeneous ASIF. It can be seen from these figures that ASIF-NPNR produces the worst results whereas ASIF-IPIR produces the best results and an ASIF generated using efficient placement and routing technique lies almost in the middle of the best and the worst solutions. The major area loss in ASIF is mainly due to the placement/partitioning inefficiency. New placement techniques need to be explored to further improve area of an ASIF.

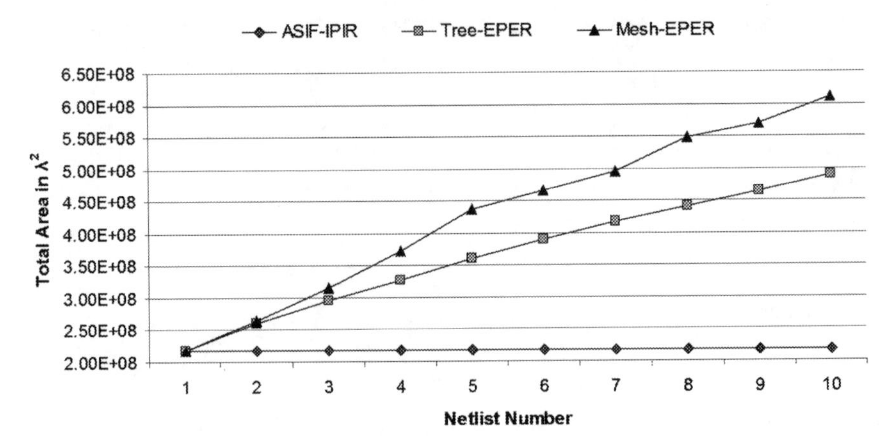

Fig. 6.15 Quality comparison between heterogeneous mesh-based and tree-based ASIF (SET I)

6.4.1 Quality Comparison Between Heterogeneous Mesh-Based and Tree-Based ASIF

In order to evaluate the quality of mesh-based and tree-based architectures, a quality comparison between heterogeneous mesh-based and tree-based ASIFs is performed. The quality comparison is performed separately for SET I and SET II benchmarks by using ASIF-EPER techniques for mesh-based and tree-based architectures and it is shown in Figs. 6.15 and 6.16 respectively. Similar to experiments in previous sub-section, ASIFs are generated for both architectures using "rs_decoder_2" and "cfft16x8" netlists and results are also compared against an ideal solution. As it can be seen from these figures that for both architectures, the quality of ASIFs of both architectures deteriorates with increase in the count of netlists. However, tree-based ASIF produces better results as compared to mesh-based ASIF. For small number of netlists, the gap between mesh-based and tree-based ASIFs is not significant but as we increase the count of netlists this gap becomes more and more significant and finally for a group of 10 netlists, tree-based ASIF is around 19%, 20% more area efficient than mesh-based ASIF for SET I and SET II benchmarks respectively.

6.5 Heterogeneous ASIF Hardware Generation

The hardware of heterogeneous ASIF is generated in a similar manner as that of homogeneous ASIF. However, modifications are performed to incorporate the effect of different types of blocks that are used by different netlists. Similar to the homogeneous ASIFs, the VHDL model generator of heterogeneous ASIFs is integrated with their exploration environments and the parameters that are supported by the exploration environment are also supported by the VHDL generator.

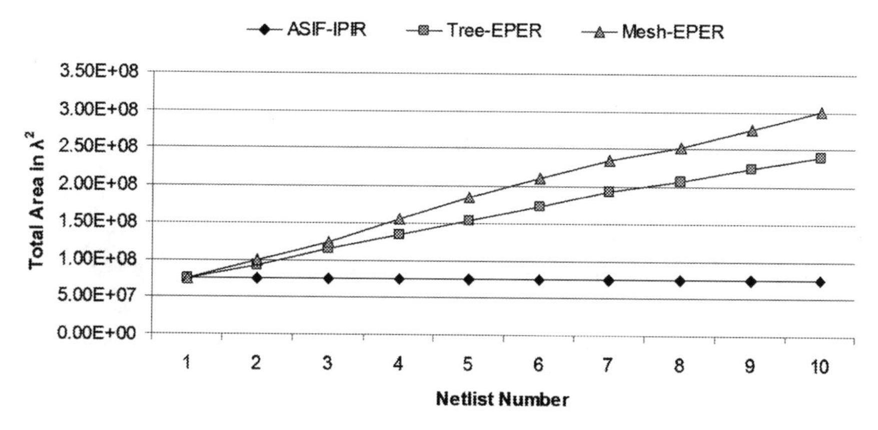

Fig. 6.16 Quality comparison between heterogeneous mesh-based and tree-based ASIF (SET II)

The ASIF generation flow remains exactly the same except that now the block database contains details about a variety of blocks that are used by different netlists being mapped on the architecture. As far as the remaining steps involved in the VHDL model generation are concerned, they remain the same. The only difference between the VHDL model generation of homogeneous and heterogeneous ASIFs is that of the support for hard-blocks. In homogeneous ASIFs all blocks are of uniform size as there is only a single type of blocks in the architecture. However, in heterogeneous architectures there are blocks of different types. Although the basic tile size remains the same, there might be blocks that occupy multiple tiles. However, it does not leave much of an effect on the the VHDL generation mechanisms. So different processes involved in the hardware generation of homogeneous ASIFs remain the same for the hardware generation of heterogeneous ASIFs.

The generated VHDL model of heterogeneous ASIF is translated to 130 nm standard cell library of STMicroelectronics, and then passed to Cadence encounter for layout generation. Some of the layout results are shown in Fig. 6.17 where tree-based ASIFs and ASICs are generated for a varying number of netlists. Figure 6.17a, c, e show the layout generated for tree-based ASIF and Fig. 6.17b, d, f show the layout of ASIC generated for 1, 5 and 9 netlists of SET II benchmarks. Here from netlist 1 we mean that layout is generated for first netlist of Table 6.2. Similarly from netlist 5 and 9, we mean that layout is generated for first 5 and all the 9 netlists of Table 6.2. The hardware of tree-based ASIF is generated using LUT-3, Arity-16 combination with maximum channel width that gives best area results. It can be seen from the figure that for 1 netlist ASIC is 20% smaller than tree-based ASIF and for 9 netlists this gap increases to almost 40%. Although for 1 netlist both ASIF and ASIC contain no switches or memories, but ASIF consumes more area than ASIC. This is because of the fact that ASIF layout is generated using a VHDL description which is the result of software flow of ASIF (refer to Fig. 4.7). During that flow different tools are used starting from the synthesis of the netlist to its routing on the architecture and these tools have added different imperfections in the design which at the end result

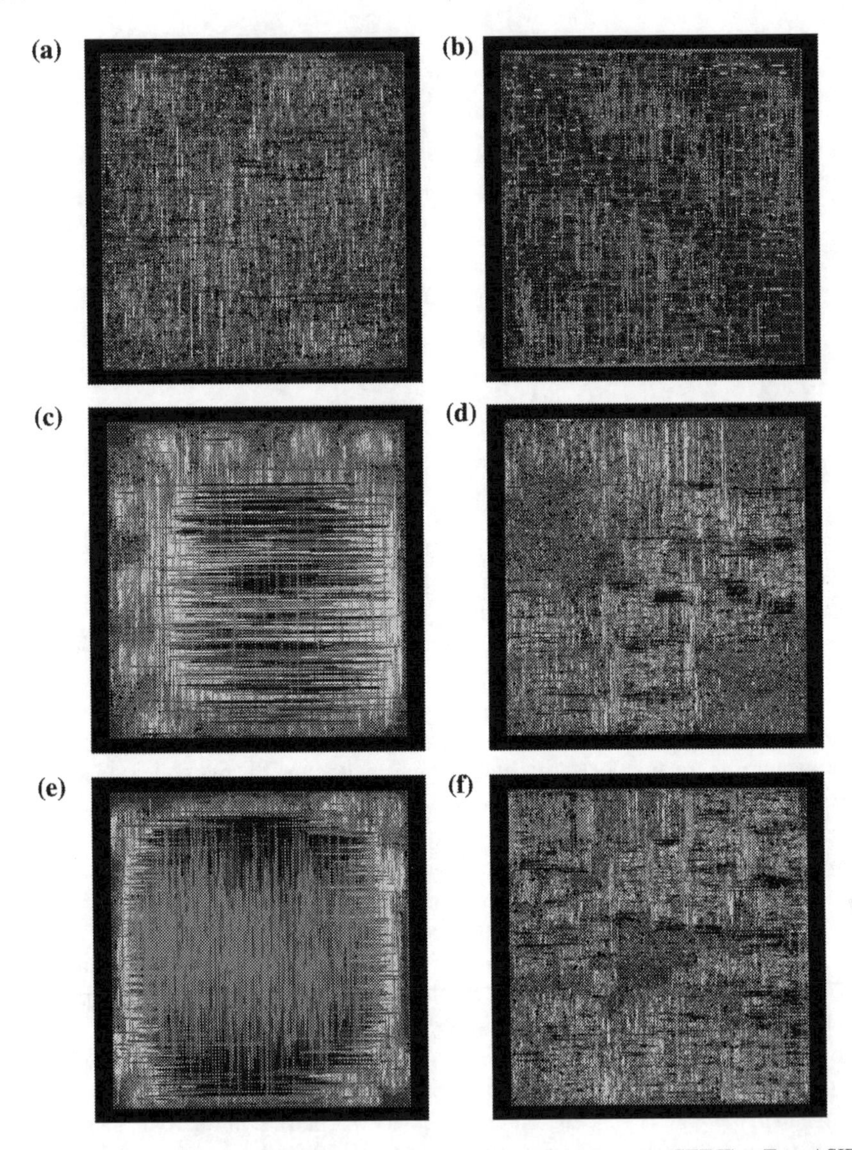

Fig. 6.17 ASIF/ASIC layout for varying number of netlists using encounter (SET II). **a** Tree-ASIF, netlists=1, 0.05 mm^2. **b** ASIC, netlists=1, 0.04 mm^2. **c** Tree-ASIF, netlists=5, 0.56 mm^2. **d** Sum of ASICs, netlists=5, 0.37 mm^2. e Tree-ASIF, netlists=9, 1 mm^2. **f** Sum of ASICs, netlists=9, 0.6 mm^2

in a larger area. Also tree-based ASIF uses LUTs which remain as configurable as in FPGAs. On the other hand, ASIC is generated directly from hardware description of the netlist and it uses gates that require no configuration bits.

Further, the increase in number of netlists increases the gap between ASIF and ASIC and for 9 netlists ASIC is 40% smaller than ASIF. This is because of the fact that increase in number of netlists increases the switches and configuration memory for ASIF while there are no configuration memories in ASICs. Also the standard cell library of STMicroelectronics does not contain SRAM cell. So LATCH cell is used instead of SRAM for configuration memory. Generally LATCH cells are larger than SRAMs and for an ASIF generated for 9 netlists, SRAMs take up to 23% of total routing area. The percentage of configuration memory area is measured from VHDL model generated using symbolic standard cell library (i.e. measure before mapping it to ST standard cell library). The high percentage of area taken by SRAMs indicates that the gap between ASIF and ASIC can further be reduced by replacing LATCH cells with SRAMS cell in ST standard cell library.

6.6 Summary and Conclusion

In this chapter, a new tree-based heterogeneous ASIF is presented. Four ASIF generation techniques are explored for 17 open core benchmarks. Based on the types of hard-blocks used, benchmarks are divided into two sets where first set contains a mixture of logic blocks and multipliers (SET I) while second set contains a mixture of logic blocks, multipliers and adders (SET II). Comparison of different techniques with tree-based FPGA shows that the most efficient ASIF generation technique is 64%, 75% more area efficient than an equivalent tree-based FPGA for SET I and SET II respectively. Further the exploration of LUT and arity size on tree-based ASIF shows that an ASIF with LUT size 4 and arity size 16 produces the most efficient results for SET I and an ASIF with LUT size 3 and arity size 16 produces the most efficient results for SET II. Comparison of tree-based ASIF with mesh-based ASIF reveals that tree-based ASIF is 11.27% better and 1.5% worse in terms of area for SET I and SET II while it consumes 69%, 70% less wires than mesh-based ASIF for two sets respectively. Finally the quality analysis of two architectures reveals that for a group of 10 similar netlists, tree-based ASIF is 19%, 20% more area efficient than mesh-based ASIF for SET I and SET II benchmarks respectively. The results presented in this chapter are also published in [115].

Layout of the tree-based heterogeneous ASIF is also performed in this chapter using Cadence encounter and its comparison with equivalent sum of ASICs shows that ASIC is 40% smaller than ASIF for a group of 9 open core netlists (SET II). However, the gap between sum of ASICs and ASIF can be reduced by replacing LATCH cells with the SRAM cells. Further, full-custom design of repeatedly used cells can be performed to further decrease the area of ASIF and reduce the gap between ASIF and ASIC. These repeatedly used cells can be groups of similar size SRAMs, switches and logic blocks etc. However, full-custom design can not be much of a benefit for ASICs as there are no major repeatedly used components in them.

Chapter 7
Conclusion and Future Lines of Research

7.1 Summary of Contributions

The main focus of this book has been the exploration and optimization of tree-based and mesh-based FPGA architectures. The study's purpose has been to find the ways to improve the overall efficiency of FPGA architecture with or without compromising their principle advantages. In this regard two distinct FPGA architectures have been under consideration: one is the island-style while the other is hierarchical. Mesh-based (island-style) architecture is a known, well explored and thoroughly investigated architecture. Tree-based (hierarchical) architecture, despite its good performance, is relatively less explored FPGA architecture. The two architectures have a lot in common in terms of basic logic and routing resources. However, the global arrangement of logic and routing resources and the detailed interconnect topologies of their switch blocks make them the two distinct architectures.

In this work a major study is carried out on the improvement of logic resource usage in heterogeneous mesh-based and tree-based FPGA architectures. For this purpose separate exploration environments are developed for the two architectures that efficiently place and route certain number of heterogeneous benchmarks on them. Although the primary objective of this work is not to establish the supremacy of one architecture over the other, yet, a detailed comparison between the two architectures is presented to highlight their advantages and disadvantages. Further, to improve the routing resource usage, an exploration of tree-based homogeneous Application Specific Inflexible FPGA (ASIF) is carried out and then its comparison with mesh-based ASIF is performed to evaluate the two ASIFs. Tree-based ASIF is then extended to heterogeneous domain where a detailed exploration of tree-based heterogeneous ASIF and its comparison with mesh-based heterogeneous ASIF is performed.

Some of the major contributions of this book are as follows:

U. Farooq et al., *Tree-Based Heterogeneous FPGA Architectures*,
DOI: 10.1007/978-1-4614-3594-5_7,

7.1.1 Heterogeneous Tree-Based FPGA Exploration Environment

Chapter 4 presented exploration environments for the exploration of heterogeneous FPGA architectures. The highlight of the work presented in Chap. 4 includes a new environment for tree-based heterogeneous FPGA architecture and an optimized environment for mesh-based heterogeneous FPGA architecture [92]. A significant amount of research work is already done regarding mesh-based heterogeneous architectures, but to the best of our knowledge all the previous work uses predetermined floor-planning technique where hard-blocks are placed in fixed columns. Although this kind of technique can be helpful for an easy and compact layout, it can lead to the wastage of precious logic and routing resources and hence increased area. This work presents an exploration environment for mesh-based FPGA that can optimize automatically the floor-planning of the FPGA architecture for a given set of applications.

A number of techniques are explored for both mesh-based and tree-based architectures using their respective environments. The techniques of the two architectures are then evaluated using the results that are obtained by mapping 21 benchmarks on the two architectures. In order to have a profound analysis of the techniques, these benchmarks are carefully selected to cover different inter-block communication trends. The results obtained after the experimentation suggest that, for a mesh-based architecture, the floor-planning technique based on the movement and rotation of logic and hard-blocks gives the best results and is much better than the one where hard-blocks are fixed in columns (i.e. the technique normally used in mesh-based architectures). For 21 benchmarks, on average, a column based floor-planning takes 19% more area, crosses 8% more switches on critical path, consumes 13% more memories and 20% more buffers than the best non-column floor-planning. Further, the comparison between different techniques of mesh-based and tree-based architecture shows that, on average, the best technique of tree-based architecture is 8.7% more area efficient, crosses 60% less switches on critical path, consumes 11% less memories and almost same number of buffers than the best non-column based technique (i.e. technique based on movement and rotation of logic and hard-blocks) of mesh-based architecture. Also, the best technique of tree-based FPGA is 22% more area efficient and crosses 62% less switches on critical path than the equivalent column-based technique of mesh-based architecture. These results are averaged for 21 benchmarks which cover different aspects of heterogeneous benchmarks.

7.1.2 Tree-Based ASIF Exploration

Chapter 5 presented a new tree-based homogeneous Application Specific Inflexible FPGAs. If a digital product is required to provide multiple functionalities at exclusive times, each distinct functionality represented by an application circuit is efficiently mapped on an FPGA. Later, unused resources of the FPGA are removed to generate an ASIF.

Conventional partitioning/placement and routing algorithms are designed for individual netlists and they do not take into account the inter-netlist optimization. In this work, however, we have modified these algorithms which has led to the exploration of four ASIF generation techniques for tree-based architecture. These techniques are divided on the basis of how each technique uses logic and routing resources of the architecture. ASIF-NPNR, first of the four techniques, uses normal partitioning/placement and routing algorithms that are employed in FPGAs and this technique in general produces the worst results among the four techniques. ASIF-EPNR employs efficient logic sharing technique to optimize the use of logic resources among different netlists but uses normal routing algorithms. This technique produces slightly better results when compared to ASIF-NPNR, but produces poor results compared to other two techniques. ASIF-NPER is the third among the four techniques. It uses no efficient logic sharing but employs efficient routing resource sharing and results in a significant improvement over the first two techniques. ASIF-EPER is the fourth and the most efficient technique as it uses both efficient logic sharing and efficient routing and produces the best over all results.

Four ASIF generation techniques are explored for a set of 16 MCNC benchmarks and comparison of these techniques with tree-based FPGA shows that the most efficient ASIF generation technique (i.e. ASIF-EPER) is 64% more area efficient than an equivalent tree-based FPGA. Further the exploration of LUT and arity size on tree-based ASIF shows that smaller LUTs with larger arity sizes give better area results whereas larger LUTs with larger arity sizes give better performance results. The area-delay product of different LUT-arity combinations reveals that an ASIF with LUT size 4 and arity size 16 produces the most efficient results.

Later the comparison between the best techniques of tree-based and mesh-based ASIFs is performed. For tree-based ASIF, LUT 4 arity 16 combination is used while for mesh-based ASIF LUT 4 is used. The two ASIFs are compared over a range of varying signal bandwidth and the results show that the best tree-based ASIF is 12%, 77% more efficient than the best mesh-based ASIF in terms of routing area and number of wires used.

The quality analysis of ASIF generation method is performed by generating an ASIF for a group of netlists for which an ideal ASIF solution is known. So, an ASIF is generated for same set of netlists. The quality analysis of tree-based ASIF reveals that ASIF-EPER produces much worse results than ideal solution. The major cause of these results is the heuristic based partitioning algorithm. New partitioning techniques need to be explored to further improve area of a tree-based ASIF. Further a quality comparison between mesh-based and tree-based ASIFs reveals that for a group of 10 similar netlists, tree-based ASIF is 33% more area efficient than mesh-based ASIF.

In Chap. 6, a new tree-based heterogeneous ASIF is presented. Similar to tree-based homogeneous ASIF, four ASIF generation techniques are explored for 17 open core benchmarks. Based on the types of hard-blocks used, benchmarks are divided into two sets where first set contains 8 benchmarks each of which contains a mixture of multipliers and logic blocks (SET I) and second set contains 9 benchmarks where each benchmark contains a mixture of multipliers and adders along with logic blocks (SET II). Comparison of different techniques with tree-based FPGA shows that the most efficient ASIF generation technique (i.e. ASIF-EPER) is 64%, 75% more area efficient than an equivalent tree-based FPGA for SET I and SET II respectively. Further the exploration of LUT and arity size on tree-based ASIF shows that an ASIF with LUT size 4 and arity size 16 produces the most efficient area-delay results for SET I and an ASIF with LUT size 3 and arity size 16 produces the most efficient area-delay results for SET II. Comparison of tree-based ASIF with mesh-based ASIF reveals that the best tree-based ASIF is 11.27% better and 1.5% worse than the best mesh-based ASIF in terms of area for SET I and SET II respectively. Further the wire comparison of mesh-based and tree-based ASIFs shows that tree-based ASIF consumes 69%, 70% less wires than mesh-based ASIF for SET I and SET II. Finally the quality analysis of two architectures reveals that for a group of 10 similar netlists, tree-based ASIF is 19%, 20% more area efficient than mesh-based ASIF for SET I and SET II benchmarks respectively.

7.1.3 FPGA and ASIF Hardware Generation for Tree-Based Architecture

Chapters 3 and 5 presented new FPGA and ASIF hardware generation models for tree-based architecture. These two chapters also included the description for hardware generation of mesh-based architecture. The FPGA and ASIF hardware generation models are similar, however, necessary modifications are made in the two models to meet the respective needs of FPGA and ASIF architectures. The hardware generator is integrated with the exploration environment. So the major benefit is that all the architecture level changes in the exploration environment are directly translated in VHDL model. The VHDL model of an FPGA and ASIF is generated for multiple netlists. Layout is performed for FPGAs and ASIFs using 130 nm 6-metal layer CMOS process of ST Microelectronics.

Chapter 6 presented the layout comparison results between tree-based heterogeneous ASIF and sum of ASICs. The results showed that for a group of 9 netlists (SET II), sum of ASICs is 40% smaller than ASIF. However, the area of ASIF is affected by the use of LATCH cells instead of SRAMs and the gap between ASIF and sum of ASICs can be reduced by replacing LATCH cells with SRAMs. Further the area of ASIF can be improved by performing full-custom design of repeatedly used cells in the architecture.

7.2 Suggestions for Future Research

In this work, we explored a number of techniques for heterogeneous tree-based and mesh-based FPGA architectures. Also we improved the density of FPGA architectures by customizing them for a pre-determined set of applications. However, there are a number of aspects that are unexplored yet and a thorough study is required to explore those aspects and in turn further improve the efficiency of FPGA architectures. Some of the suggestions for future research are described below. These suggestions are applicable to both mesh-based and tree-based architectures unless otherwise specified.

7.2.1 Datapath Oriented FPGA Architectures

During past few years FPGAs have seen a rapid growth in their logic capacity which has led to the increasing use of FPGAs for the implementation of arithmetic-intensive applications. Arithmetic-intensive applications often contain large portion of datapath circuits. Datapath circuits usually contain hard-blocks (e.g. multipliers, adders, memories etc.) that are connected together by regularly structured signals called buses. Conventional FPGAs do not use the regularity of datapath circuits. So it is possible to modify the conventional FPGA architectures to exploit the regularity of datapath circuits and achieve significant area savings.

Although some work has been done in this regard for mesh-based homogeneous architecture [127], but to the best of our knowledge no work has been done yet in heterogeneous domain neither for mesh-based nor for tree-based architecture. Both heterogeneous mesh-based and tree-based FPGA architectures can be modified to exploit the regularity of datapath circuits and improve the area efficiency. Generalized proposition of a datapath oriented tree-based architecture is shown in Fig. 7.1. Contrary to the conventional one, the proposed version could be divided into two sub-structures. One sub-structure contains CLBs while other sub-structure contains only HBs. Interconnect between the two sub-structures and inside the sub-structure containing only CLBs is fine-grain while it is coarse-grain (i.e. bus-based) in the sub-structure containing only HBs. Main motivation behind the migration from single structure to two sub-structures is:

1. Most of the arithmetic-intensive applications have two parts: datapath part and the control part. Control part is implemented using CLBs while datapath part is implemented using HBs. So it is better to manage them separately due to their different communication behaviors.
2. Division of single structure into two sub-structures helps optimizing their logic and routing resources independently; thus leading to better logic and routing density of the architecture.

Preliminary experimentation has shown some promising results and for a group of four datapath circuits, the datapath oriented tree-based architecture consumes

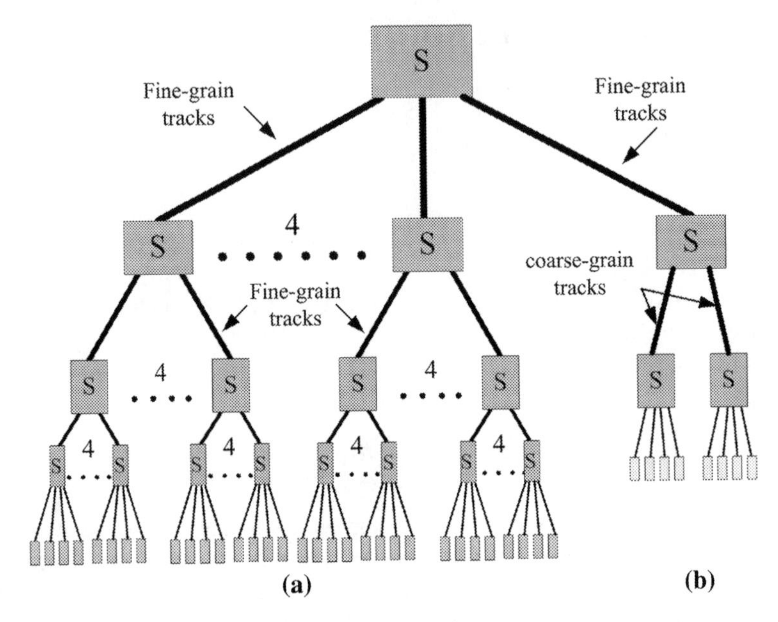

Fig. 7.1 Generalized example of datapath oriented tree-based FPGA architecture. **a** Sub-structure containing only CLBs. **b** Sub-structure containing only HBs

21% less area than the conventional tree-based heterogeneous FPGA architecture. However, these are only preliminary results as the benchmarks are small both in size and in number and improvements regarding resource partitioning are required to further enhance the results.

As far as the mesh-based heterogeneous FPGA architecture is concerned, modification can be made in it in a manner similar to the one suggested by [127]. However, there might be few exceptions as the architecture proposed by [127] uses only homogeneous blocks.

7.2.2 Timing Analysis

One of the major future work is to perform accurate timing analysis of both FPGA and ASIFs. In this work we presented timing results based on the count of critical path switch number. Although this is not accurate as it does not contain the wire delays, it gives an idea about the timing behavior of the architecture. In future we want to perform the complete timing analysis of both mesh-based and tree-based architecture. This timing analysis will include measuring the critical path, optimizing the critical path using timing driven routing, and finding an accurate compromise between area and delay of an architecture.

7.2.3 Integrating ASIF Blocks in an FPGA Architecture

Commercial FPGA vendors use different variety of hard-blocks in their FPGAs. They provide a range of FPGA device variants to fulfill varying domain-specific requirements of their customers. Smaller ASIF blocks can also be integrated in FPGAs to enhance the domain-specific needs of customers. An ASIF block serves as a multi-tasking hard-block that can support a number of different functionalities at mutually exclusive times. ASIF blocks can be designed for some general purpose DSP requirements, or for more specific applications such as video decoders/encoders applications etc.

7.2.4 Further Optimizing the ASIF Generation

In this work a new tree-based ASIF is presented and its comparison with equivalent tree-based FPGA shows that it gives significantly better area results than tree-based FPGA. Tree-based ASIF is basically an intermediate solution between FPGAs and ASICs where a certain amount of flexibility is retained while removing unnecessary resources. Although the removal of unnecessary resources in ASIF gives better area than FPGA, it makes the structure irregular; hence making the layout of ASIF more difficult than FPGA. So from this point on, the possible research directions in ASIF architecture can be as follows (these directions are general in nature and they are applicable to both mesh-based and tree-based ASIFs unless otherwise specified):

- One future direction for ASIF can be to improve their efficiency by further optimizing their logic and routing resources. In Chap. 6, the comparison between tree-based ASIF and Sum of ASICs revealed that tree-based ASIF consumes more area than Sum of ASICs. This is because of the fact that tree-based ASIF uses LUTs and they remain as configurable as in FPGA. So, the possible optimizations in the ASIF generation can be as follows:

1. Different set of application netlists, mapped on an ASIF, program the SRAM bits of a LUT differently. If all the netlists program a particular SRAM of a LUT in a similar fashion, that SRAM bit can be replaced by a hard-coded 0 or 1 eventually leading to further improvement in the area of ASIF. Similarly, if a 2-input multiplexor in a LUT receives similar hard-coded values, that multiplexor can be removed, and replaced by the hard-coded value. A major element of research is to tailor the LUT configuration of different netlists in such a way that maximum logic block optimization is achieved.
2. Also, the routing network of ASIFs contains SRAMs that are used to provide limited configurability. However, this configurability is not much of a use as tools do not guaranty the mapping of extra circuits. So, the routing network of ASIF can be further optimized as there is a good probability that certain SRAMs of the routing network have same value over all the circuits that are required

to be implemented on ASIF. In such a case these SRAMs can be replaced with constants; hence resulting in further improvement of area results.

- The generation of ASIF is based on the removal of unused resources which makes it irregular and tile-based layout of ASIF does not remain possible. So, another future direction on ASIF can be to add more flexibility to it than required. The main objective of this additional flexibility would be to have a minimum tile or a set of tiles that can be repeated to perform the layout of ASIF. Although additional flexibility will reduce the area gain of ASIF compared to FPGA, it will make the layout easier; hence resulting in faster time to market. Further, it will increase the probability of implementing additional circuits that are not in the set for whom the ASIF is generated.
- Quality analysis of ASIF in Chaps. 5 and 6 revealed that there is significant room for improvement in the placement/partitioning algorithms. This improvement can be achieved by better exploiting the inter-netlist dependance. However, in this approach, a major element of research would be to optimize the inter-netlist dependance without compromising the intra-netlist optimization.
- In Chap. 6, area comparison between ASIF and sum of ASICs revealed that even for 1 netlist ASIC is 20% smaller than ASIF. This is because of the fact that in case of ASIFs, the netlist synthesis involves a number of open source tools. These tools are not optimal and they add some imperfections while converting the netlist from its hardware description to .net format. On the contrary, ASIC generation involves only one commercial synthesis tool. So, the deficiencies introduced by open source tools play an important role in the area gap between ASIF and sum of ASICs and this gap can be reduced by performing the netlist synthesis using more efficient commercial tools.

The above mentioned optimizations an be combined together to give either an ASIF with better area density; hence reduced gap between ASIF and ASIC or an ASIF with reduced layout efforts; hence faster time to market.

7.2.5 The Unexplored Parameters of Mesh-Based Architecture

In this work we have used single length, unidirectional routing network with CLB size 1 for mesh-based architecture which is at the core of both FPGA architecture and ASIF architecture. However, it would be interesting to explore the behavior of mesh-based ASIF having CLB size greater than 1, or having a mixture of different length routing wires some of whom span a single architecture tile whereas others span multiple tiles or having a mixture of fine-grain and coarse-grain routing wires.

References

1. Aggarwal, A., Lewis, D.M.: Routing architecture for hierarchical field programmable gate arrays. In: International Conference on Computer Design, pp. 475–478 (1994)
2. Actel: Proasic3 flash family FPGAs. http://www.actel.com/documents/pa3-ds.pdf (Oct 2005)
3. DeHon, A.: Reconfigurable architectures for general purpose computing. AI Technical Report1586, MIT Artificial Intelligence Laboratory (Sept 1996)
4. DeHon, A.: Balancing interconnect and computation in a reconfigurable computing array (or, why you don't really want 100% LUT utilization). In: International Symposium on Field Programmable Array FPGA, Monterey, CA, pp. 69–78 (Feb 1999)
5. DeHon, A.: Rent's rule based switching requirements. System Level Interconnect Prediction Workshop (2001)
6. Dunlop, A., Kernighan, B.: A procedure for placement of standard-cell VLSI circuits. IEEE Trans. CAD, pp. 92–98 (Jan 1985)
7. Ahmed, E., Rose, J.: The effect of LUT and cluster size on deep-submicron FPGA performance and density. In: Proceedings of the International Symposium on Field Programmable Gate Arrays, pp. 3–12 (2000)
8. Ahmed, E., Rose, J.: The effect of LUT and cluster size on deep-submicron FPGA performance and density. IEEE (2003)
9. Alliance: http://www-asim.lip6.fr/recherche/alliance/ (2006)
10. Altera: Flex 10k embedded programmable logic device family, ds-f10k-4.2. http://www.altera.com/literature/ds/dsf10k.pdf (Jan 2003)
11. Altera: http://www.altera.com (2010)
12. Altera: 40-nm FPGA power management and advantages. http://www.altera.com (2010)
13. Altera: Altera. http://www.altera.com (2010)
14. Marquart, A., Betz, V., Rose, J.: Using cluster-based logic block and timing-driven packing to improve FPGA speed and density. In: International Symposium on FPGA, Monterey, pp. 37–46 (1999)
15. Apex, A.: Apex 20k programmable logic device family data sheet, ds-apex20k-5.1. http://www.altera.com/literature/ds/apex.pdf (March 2000)
16. ApexII, A.: Apex ii programmable logic device family, dsapexii-3.0. http://www.altera.com/literature/ds/dsap2.pdf (Aug 2002)
17. Singh, A., Marek-Sadowska, M.: Efficient circuit clustering for area and power reduction in FPGAs. In: International Symposium on Field Programmable Gate Arrays, pp. 59–66 (2002)
18. ATMEL: http://www.atmel.com (2010)

U. Farooq et al., *Tree-based Heterogeneous FPGA Architectures*,
DOI: 10.1007/978-1-4614-3594-5,

19. Beauchamp, M., Hauck, S., Underwood, K., Hemmert, K.: Embedded floating-point units in FPGAs. In: FPGA, pp. 12–20 (2006)
20. Berkeley: Berkeley logic interchange format (BLIF). http://vlsi.colorado.edu/vis/blif.ps (1992)
21. Berkeley: ABC: A system for sequential synthesis and verification. http://www.eecs.berkeley.edu/alanmi/abc/ (2007)
22. Betz, V., Marquardt, A., Rose, J.: Architecture and CAD for Deep-Submicron FPGAs. Kluwer Academic Publishers, New York (Jan 1999)
23. Birkner, J., Chan, A., Chua, H., Chao, A., Gordon, K., Kleinman, B., Kolze, P., Wong, R.: Avery-high-speed field-programmable gate array using metal-to-metal antifuse programmable elements. Microelectron. J. **23**(7), 561–568 (1992)
24. Kernighan, B., Lin, S.: An efficient heuristic procedure for partitioning graphs. Bell Syst. Tech. J. **49**, 291–307 (1970)
25. Landman, B., Russo, R.: On pin versus block relationship for partition of logic circuits. IEEE Trans. Comput. **20**, 1469–1479 (1971)
26. Brayton, R., Hachtel, G., Sangiovanni-Vincentelli, A.: Multilevel logic synthesis. Proc. IEEE **78**(2), 264–300 (Feb 1990)
27. Brayton, R., McMullen, C.: The decomposition and factorization of Boolean expressions.Proc. ISCAS, 29–54 (1982)
28. Callahan, T.J., Hauser, J.R.,Wawrzynek, J.: The garp architecture and C compiler. Computer **33**(4), 62–69 (April 2000)
29. Cherepacha, D., Lewis, D.: DP-FPGA: an FPGA architecture optimized for datapaths. VLSI Des. **4**(4), 329–343 (1996)
30. Ho, C.H., Leong, P.H.W., Luk, W., Wilton, S., Lopez-Buedo, S.: Virtual embedded blocks: a methodology for evaluating embedded elements in FPGAs. In: FCMM, 35–44 (2006)
31. Alpert, C.J., Hagen, L.W., Kahng, A.B.: Multilevel circuit partitioning. In: Design Automation Conference, pp. 530–533 (1997)
32. Alpert, C.J., Chan, T., Huang, D., Kahng, A., Markov, I., Mulet, P., Yan, K.: Faster minimization of linear wirelength for global placement. In: ACM Symposium on Physical Design, pp. 4–11 (1997)
33. Leiserson C.: Fat-trees: universal networks for hardware efficient supercomputing. IEEE Trans. Comput. **C34**(10), 892–901 (Oct 1985)
34. Fiduccia, C.M., Mattheyeses, R.M.: A linear-time heuristic for improving network partitions. In: Design Automation Conference, pp. 175–181 (1982)
35. Compton, K., Hauck, S.: Automatic design of area-efficient configurable ASIC cores. IEEE Trans. Comput. 56(5), 662–672 (May 2007)
36. Open core: http://www.opencores.org/ (2009)
37. Sechen, C., Sangiovanni-Vincentelli, A.: The timberwolf placement and routing package. JSSC, 510–522 (April 1985)
38. Papa, D.A., Markov, I.L.: Hypergraph partitioning and clustering. Technical Report, University of Michigan, EECS Department (1987)
39. Huang, D., Kahng, A.: When clusters meet partitions: new density based methods for circuit decomposition. In: IEEE European Design and Test Conference, pp. 60–64 (1995)
40. Huang, D., Kahng, A.: Partitioning-based standard-cell global placement with an exact objective. In: ACM Symposium on Physical Design, pp. 18–25 (1997)
41. eASIC: http://www.easic.com (2010)
42. Ebeling, C., Cronquist, D.C., Franklin, P.: RaPiD-Reconfigurable pipelined datapath. In: Field Programmable Logic and Applications, pp. 126–135 (1996)
43. Bozorgzadeh, E., et al.: Routability-driven packing: metrics and algorithms for cluster-based FPGAs. IEEE J. Circuits Syst. Comput. **13**(1), 77–100 (2004)
44. El Gamal, A., Greene, J., Reyneri, J., Rogoyski, E., El-Ayat, K., Mohsen, A.: An architecture for electrically configurable gate arrays. IEEE J. Solid-State Circuits **24**(2), 394–398 (April 1989)

45. Farooq, U., Parvez, H., Marrakchi, Z., Mehrez, H.: A new tree-based coarse-grained FPGA architecture. Research in Microelectronics and Electronics, 2009. PRIME 2009. Ph.D, pp. 48–51 (2009)
46. Farooq, U., Marrakchi, Z., Mrabet, H., Mehrez, H.: The effect of lut and cluster size on a tree based FGPA architecture. In: Proceedings of the 2008 International Conference on Reconfigurable Computing and FPGAs, pp. 115–120. IEEE Computer Society, Washington, DC, USA (2008). http://portal.acm.org/citation.cfm?id=1494647.1495190
47. Fiduccia, C.M., Mattheyeses, R.M.: A linear-time heuristic for improving network partitions. Proc. DAC, pp. 175–181 (1982)
48. Karypis, G., Aggarwal, R., Kumar, V., Shekhar, S.: Multilevel hypergraph partitioning: applicationin VLSI design. In: Design Automation Conference, pp. 526–529 (1997)
49. Karypis, G., Kumar, V. : Multilevel k-way hypergraph partitioning. In: Design Automation Conference (1999)
50. Karypis, G., Kumar, V.: Multilevel k-way hypergraph partitioning. In: Design Automation Conference (1999)
51. Lemieux, G., Lee, E., Tom, M., Yu, A.: Directional and single-driver wires in FPGA interconnect. In: IEEE International Conference on Field-Programmable Technology (ICFPT), pp. 41–48 (2004)
52. Govindu, G., Choi, S., Prasanna, V., Daga, V., Gangadharpalli, S., Sridhar, V.: A high performance and energy-efficient architecture for floating-point based LU decomposition on FPGAs. In: Proceedings of the 18th International Parallel and Distributed Processing Symposium (2004)
53. Sigl, G., Doll, K., Johannes, F.: Analytical placement: a linear or a quadratic objective function? In: Design Automation Conference, pp. 427–432 (1991)
54. Guterman, D.C., Rimawi, I.H., Chiu, T.L., Halvorson, R., McElroy, D.: An electrically alterable nonvolatile memory cell using a floating-gate structure. IEEE Trans. Electron Devices **26**(4), 576–586 (April 1979)
55. Hamdy, E., McCollum, J., Chen, S., Chiang, S., Eltoukhy, S., Chang, J., Speers, T., Mohsen, A.: Dielectric based antifuse for logic and memory ICs. In: IEEE International Electron Devices Meeting, IEDM'88, Technical Digest, pp. 786–789 (1988)
56. HardCopy: HardCopy IV ASICs, device handbook. Available at http://www.altera.com/products/devices/hardcopy-asics/hardcopy-iv/literature/hcivliterature.jsp (IV)
57. Hauck, S., Burns, S., Borriello, G., Ebeling, C.: An FPGA Implement. Asynchronous Circuits. IEEE Des. Test **11**(3), 60–69 (1994)
58. Ho, C., Leong, P., Luk, W., Wilton, S., Lopez-Buedo, S.: Virtual embedded blocks: a methodology for evaluating embedded elements in FPGAs. In: Proceedings of the FCCM, pp. 35–44 (2006)
59. Hutton, M., Yuan, R., Schleicher, J., Baeckler, G., Cheung, S., Chua, K., Phoon, H.: A methodology for FPGA to structured-ASIC synthesis and verification. DATE **2**, 64–69 (March 2006)
60. Kuon, I., Rose, J.: Measuring the gap between FPGAs and ASICs. IEEE Trans. CAD **26**(2), 203–215 (2007)
61. Birkner, J., Chan, A., Chua, H.T., Chao, A., Gordon, K., Kleinman, B., Kolze, P., Wong, R.: A very-high-speed field programmable gate array using metal-to-metal antifuse programmable elements. Micro **23**(7), 561–568 (Nov 1992)
62. Jacomme, L.: VHDL analyser for synthesis (vasy) asim/lip6/upmc (2000)
63. Jamieson, P., Luk, W., Wilton, S., Constantinides, G.: An Energy and Power Consumption Analysis of FPGA Routing Architectures, pp. 324–327 (2009)
64. Cong, J., Ding, Y.: Flowmap: an optimal technology mapping algorithm for delay optimization in lookup-table based FPGA Designs. IEEE Trans. CAD, pp. 1–12 (1994)
65. Cong, J., Ding, Y.: On area/depth trade-off in LUT-based FPGA technology mapping. IEEE Trans. VLSI Syst. **2**(2), 137–148 (1994)

66. Cong, J., Ding, Y.: Structural gate decomposition for depth-optimal technology in LUT-based FPGA designs. ACM Trans. Des. Autom. Electron. Syst. **5**(3) (2000)
67. Cong, J., Hwang, Y.: Simultaneous depth and area minimization in LUT-based FPGA mapping. In: ACM/SIGDA International Symposium on Field Programmable Gate Array, pp. 68–74 (1995)
68. Frankle, J.: Iterative and adaptive slack allocation for performance-driven layout and FPGA routing. In: ACM/IEEE Design Automation Conference, pp. 536–542 (1992)
69. Jia, X., Vemuri, R.: Studying a GALS FPGA Architecture Using a Parameterized Automatic Design Flow, pp. 688–693 (2006)
70. Jones, A.K., Hoare, R., Kusic, D., Fazekas, J., Foster, J.: An FPGA-based VLIW processor with custom hardware execution. In: Proceedings of the International Symposium on Field Programmable Gate Arrays, pp. 107–117 (2005)
71. Rose, J., Francis, R., Lewis, D., Chow, P.: Architecture of field-programmable gate arrays: the effect of logic functionality on area efficiency. IEEE J. Solid State Circuits **25**(5), 1217–1225 (Oct 1990)
72. Kuon, I., Rose, J.: Measuring the gap between FPGAs and ASICs. In: Proceedings of the ACM/SIGDA 14th International Symposium on Field Programmable Gate Arrays, pp. 21–30, ACM New York, NY (2006)
73. Lagadec, L.: Abstraction, modlisation et outils de CAO pour les architectures reconfigurables. Ph.D. thesis, Universit de Rennes 1 (2000)
74. Landman, B., Russo, R.: On pin versus block relationship for partition of logic circuits. IEEE Trans. Comput. **20,** 1469-1479 (1971)
75. Sanchis, L.A.: Multiple-way network partitioning. IEEE Trans. Comput. **38**(1),62–81 (1989)
76. Lattice: Latticeecp/ec family data sheet versio 02.0. http://www.latticesemi.com/lit/docs/datasheets/fpga/ecp_ec_datasheet.pdf (Sept 2005)
77. Lemieux, G., Lee, E., Tom, M., Yu, A.: Directional and single-driver wires in FPGA interconnect. In: IEEE International Conference on Field-Programmable Technology (ICFPT), pp.41–48 (2004)
78. Hagen, L., Kahng, A.: Combining problem reduction and adaptive multi-start: a new technique for superior iterative partitioning. In: IEEE Trans. CAD, pp.92–98 (1997)
79. Lip6: http://www-asim.lip6.fr (2007)
80. McMurchie, L., Ebeling, C.: Pathfinder: a negotiation-based performance-driven router for FPGAs. In: International Workshop on Field Programmable Gate Array, pp. 111–117 (1995)
81. Luu, J., Kuon, I., Jamieson, P., Campbell, T., Ye, A., Fang, W.M., Rose, J.: VPR 5.0: FPGACAD and architecture exploration tools with single-driver routing, heterogeneity and process scaling. In: FPGA, pp. 133–142 (Feb 2009)
82. Marrakchi, Z.: Exploration and optimization of tree-based FPGA architectures. Ph.D. thesis: http://www-asim.lip6.fr/publications/ (2008)
83. Marshall, A., Stansfield, T., Kostarnov, I., Vuillemin, J., Hutchings, B.: A reconfigurable arithmetic array for multimedia applications. In: International Symposium on Field Programmable Gate Arrays, pp. 135–143 (1999)
84. Dehkordi, M., Brown, S.: The effect of cluster packing and node duplication control in delay driven clustering. In: IEEE International Conference on Field Programmable Technology (ICFPT), pp. 227–233 (2002)
85. Garey, M., Johnson, D.: Computers and Intractability: A Guide to the Theory of NP Completeness. Freeman, San Francisco, CA (1979)
86. Huang, M., Romeo, F., Sangiovanni-Vincentelli, A.: An efficient general cooling schedule for simulated annealing. In: ICCAD, pp. 381–384 (1986)
87. Hutton, M., Adibsamii, K., Leaver, A.: Timing-driven placement for hierarchical programmable logic devices. In: International Symposium on Field Programmable Gate Array, pp. 3–11 (2001)

88. MicroBlaze: MicroBlaze soft processor core. Available at http://www.xilinx.com/tools/microblaze.htm (2010)
89. Miyamoto, N., Ohmi, T.: Delay evaluation of 90 nm CMOS multi-context FPGA with shift-register-type temporal communication module for large-scale circuit emulation. In: IEEE International Conference on Field Programmable Technology (ICFPT), pp. 365–368 (2008)
90. NIOS: NIOS II processor. Available at http://www.altera.com/products/ip/processors/nios2/ni2-index.html (II)
91. Okamoto, T., Kimoto, T., Maeda, N.: Design methodology and tools for NEC electronics structured ASIC. In: Proceedings of the ISPD, pp. 90–96 (April 2004)
92. Parvez, H., Marrakchi, Z., Farooq, U., Mehrez, H.: A new coarse-grained FPGA architecture exploration environment. In: IEEE International Conference on ICECE Technology. FPT 2008, pp. 285–288 (2008)
93. Parvez, H., Marrakchi, Z., Mehrez, H.: ASIF: Application specific inflexible FPGA. In: International Conference on Field-Programmable Technology (FPT), pp. 112–119 (2009)
94. Parvez, H.: Design and exploration of application specific mesh-based heterogeneous FPGA architectures. Ph.D. thesis, University Pierre et Marie Curie, Paris, France (2010)
95. Du, P., Grewal, G.W., Areibi, S., Banerji, D.K.: A Fast Hierarchical Approach to FPGA Placement. In: ESA/VLSI, pp. 497–503 (2004)
96. Peter Yiannacouras, J.G.S., Rose, J.: Exploration and Customization of FPGA-based soft processors. IEEE Trans. Comput. Aided Des. Integr. Circuits Syst. **26**(2), 266–277 (Feb 2007)
97. Guerrier, P., Greiner, A.: A generic architecture for on chip packet-switched interconnections. In: Proceedings of the Design Automation and Test in Europe Conference 2000 (DATE 2000), Paris, France pp. 250–256 (Mar 2000)
98. Pistorius, J., Hutton, M., Schleicher, J., Iotov, M., Julias, E., Tharmalignam, K.: Equivalence verification of FPGA and structured ASIC implementations. In: FPL'07, pp. 423–428 (Aug 2007)
99. Hitchcock, R., Smith, G., Cheng, D.: Timing analysis of computer-hardware. IBM J. Res. Dev. 100–105 (Jan 1983)
100. Murgai, R., Brayton, R., Sangiovanni-Vincentelli, A.: On clustering for minimum delay/area. In: IEEE International Conference on Computer Aided Design, pp. 6–9 (1991)
101. Semiconductor, P.: Era60100 preliminary data sheet (1989)
102. Sentovich, E.M. et al.: Sis: a system for sequential circuit analysis. Tech. Report No. UCB/ERLM92/41, University of California, Berkeley (1992)
103. Sherlekar, D.: Design considerations for regular fabrics. In: Procedings of the ISPD, pp. 97–102 (April 2004)
104. Sima, M., Cotofana, S., van Eijndhoven, J.T.J., Vassiliadis, S., Vissers, K.: An 8x8 IDCT implementation on an FPGA-augmented trimedia. In: Proceedings of the 9th Annual IEEE Symposium on Field Programmable Custom Computing Machines, pp. 160–169 (2001)
105. Kirkpatrick, S., Gelatt, C.D., Vecchi, M.P.: Optimization by simulated annealing. Science **220**, 671–680 (1983)
106. Stratix: Stratix II FPGAs. Available at http://www.altera.com/products/devices/stratix2/st2-index.jsp (II)
107. Stratix: Stratix IV FPGAs, device handbook. Available at http://www.altera.com/products/devices/stratix-fpgas/stratix-iv/literature/stiv-literature.jsp (IV)
108. Yang, S.: Logic Synthesis and Optimization Benchmarks, Version 3.0. Microelectronics Center of North Carolina (MCNC), Raleigh (1991)
109. Tabula: http://www.tabula.com (2010)
110. Bui, T., Chaudhuri, S., Leighton, T., Sipser, M.: Graph Bisection Algorithms with Good Average Behavior. Combinatorica (1987)
111. Cormen, T., Leiserson, C., Rivest, R.: Introduction to Algorithms. MIT Press, Cambridge (1990)

112. Teifel, J., Manohar, R.: An asynchronous dataflow FPGA architecture. IEEE Trans. Comput. **53**(11), 1376–1392 (2004)
113. Tierlogic: http://www.tierlogic.com (2010)
114. Trimberger, S., Carberry, D., Johnson, A., Wong, J.: A Time-Multiplexed FPGA. pp. 22–28 (1997)
115. Farooq, U., Husain Parvez, Z.M., Mehrez, H.: Comparison between heterogeneous meshbased and tree-based application specific FPGA. In: 7th International Symposium on Applied Reconfigurable Computing (ARC'11), pp. 218–219 (2011)
116. Farooq, U., Husain Parvez, Z.M., Mehrez, H.: Exploration of heterogeneous FPGA architectures. Int. J. Reconfigurable Comput. 18 (2011)
117. Farooq, U., Husain Parvez, Z.M., Mehrez, H.: Exploring the effect of lut and arity size on a tree-based application specific inflexible FPGA. In: IEEE Conference on Design and Technology in Nanoscale Era (DTIS) (2011)
118. Underwood, K., Hemmert, K.: Closing the gap: CPU and FPGA trends in sustainable floating-point BLAS performance. In: 12th Annual IEEE Symposium on Field-Programmable Custom Computing Machines, 2004. FCCM 2004, pp. 219–228 (2004)
119. Toronto University: http://www.eecg.utoronto.ca/vpr/ (2009)
120. Betz, V., Rose, J.: VPR: a new packing placement and routing tool for FPGA Research. In: International Workshop on FPGA, pp. 213–222 (1997)
121. Carter, W., Duong, K., Freeman, R., Sze, S.: A user programmable reconfiguration gate array. In: IEEE Custom Integrated Circuits Conference, pp. 233–235 (May 1986)
122. Carter, W., Duong, K., Freeman, R.H., Hsieh, H., Ja, Y.J, Mahoney, J.E., Ngo, L T., Sze, S.L.: A user programmable reconfiguration gate array. In: IEEE Custom Integrated Circuits Conference, pp. 233–235 (May 1986)
123. Wilton, S.: Architectures and algorithms for field-programmable gate arrays with embedded memory. Ph.D. thesis, Citeseer (1997)
124. Wong, S., So, H., Ou, J., Costello, J.: A 5000-gate CMOS EPLD with multiple logic and interconnect arrays. In: Proceedings of the IEEE Custom Integrated Circuits Conference (1989), pp. 5–8 (2002)
125. Wu, K., Tsai, Y.: Structured ASIC, evolution or revolution. In: Proceedings of the ISPD, pp. 103–106 (April 2004)
126. Xilinx: Xilinx. http://www.xilinx.com (2010)
127. Ye, A., Rose, J., Lewis, D.: Architecture of datapath-oriented coarse-grain logic and routing for FPGAs, pp. 61–66 (2003)
128. Ye, A.G., Rose, J.: Using bus-based connections to improve field-programmable gate-array density for implementing datapath circuits. IEEE Trans. Very Large Scale Integr. (VLSI) Syst. **14**(5), 462–473 (May 2006)
129. Lay, Y., Wang, P.: Hierarchical interconnection structures for field programmable gate arrays. IEEE Trans. VLSI Syst. **5**(2), 186–196 (1997)
130. Sanker, Y., Rose, J.: Trading quality for compile time: ultra-fast placement for FPGAs. In: International FPGA symposium (1999)
131. Zakaria, H.: Asynchronous architecture for power efficiency and yield enhancement in the decananometric technologies: application to a multi-core system-on-chip. Ph.D. thesis, TIMA, Grenoble France (2011)
132. Zied, M., Hayder, M., Umer, F., Habib, M.: FPGA interconnect topologies exploration. Internat. J. Reconfigurable Comput. (2009)
133. Marrakchi, Z., Mrabet, H., Mehrez, H.: Hierarchical FPGA clustering to improve routability. In: Conference on Ph.D Research in Microelectronics and Electronics, PRIME (2005)